Palgrave Studies in Environmental Transformation, Transition and Accountability

Series Editor
Beth Edmondson
School of Arts
Federation University
VIC, Australia

The monographs and edited collections published in this series will be unified by interdisciplinary scholarship that considers and interrogates new knowledge of opportunities for sustainable human societies through environmental transformations, transitions and accountabilities.

These publications will integrate theoretical debates and perspectives in the natural and social sciences with sustained and detailed analysis of local, regional and international initiatives responding to environmentally driven imperatives such as climate change, fresh water, energy resources, food security, and biodiversity.

More information about this series at
http://www.palgrave.com/gp/series/15884

Beth Edmondson · Stuart Levy
Editors

Transformative Climates and Accountable Governance

Editors
Beth Edmondson
School of Arts
Federation University
Mount Helen, VIC, Australia

Stuart Levy
Federation College
Federation University
Churchill, VIC, Australia

ISSN 2523-8183 ISSN 2523-8191 (electronic)
Palgrave Studies in Environmental Transformation, Transition and Accountability
ISBN 978-3-030-07350-3 ISBN 978-3-319-97400-2 (eBook)
https://doi.org/10.1007/978-3-319-97400-2

Acknowledgements

Undertaking a book to completion is a time-consuming undertaking that invariably impacts the lives of many people. Rachael Ballard provided us with timely warmth, clarity, and a balanced sense of possibilities that made more difference than she knew. We extend our thanks to family, friends, and colleagues who have provided support and encouragement, especially to those who have been through this with us on previous occasions. We also extend our thanks to our collaborating and generous colleagues who brought their scholarship and wisdom into the chapters they wrote for this collection. We are grateful to you for reminding us that wicked problems can be solved, solutions can be scaled, and accountability is a foundation of integrity. Thank you all. We hope that you will find this book to have been worth your patience.

—Beth Edmondson and Stuart Levy

We would like to thank the editors, Beth Edmondson and Stuart Levy, for the opportunity to contribute to this text, and for thus enabling us to further explore how our own differing commitments to theoretical and empirical climate change research can be interwoven.

—Josephine Mummery and Jane Mummery

This research was supported by an Australian Research Council Future Fellowship on *Intellectual Property and Climate Change: Inventing Clean Technologies*.

—Matthew Rimmer

An earlier version of this chapter was presented at the 2016 American Political Science Association Annual Meeting. I thank the panel discussant, Sheri Breen, and chair, Michael Lipscomb, as well as the audience, for their comments on that occasion. I am also grateful to Glen Peters, Steven J. Davis, and Ken Caldeira for sharing data that were central to this chapter's arguments. Finally, I thank George Klosko, Colin Bird, Jennifer Rubenstein, Matthew Adams, Jordanna Faye Brown, and this collection's two outstanding editors—Beth Edmondson and Stuart Levy—for their invaluable feedback on earlier drafts.

—Ross Mittiga

This research was supported by an Australian Research Council Discovery Grant, *Ethics, Responsibility and the Carbon Budget*. The chapter also benefitted from comments from Alex Lenferna.

—Jeremy Moss

Contents

Notes on Contributors

Hugh Breakey is a Senior Research Fellow in moral philosophy at Griffith University. His research spans political theory, normative ethics, governance studies, and applied philosophy, exploring the ethical challenges in such diverse fields as peacekeeping, institutional governance, climate change, sustainable tourism, private property, professional ethics, and international law.

Timothy Cadman is a Research Fellow at Griffith University and specializes in the governance of sustainable development, environmental politics and policy, climate change, natural resource management including forestry, responsible investment, and institutional performance.

Jack DeWaard is an Assistant Professor in the Department of Sociology and Graduate Faculty in Population Studies in the Minnesota Population Center at the University of Minnesota. As a sociologist and demographer strongly committed to interdisciplinary research and collaborations, he studies the causes, characteristics, and consequences of human migration in and outside of the USA.

Beth Edmondson is a Senior Lecturer in the School of Arts at Federation University Australia. Her research focuses on international responses to global climate change, the possibilities for order in the international political system, the nature of sovereignty, and the scope of international law in constructing governmental capacities.

Martina Grecequet is a Postdoctoral Researcher at the Institute on the Environment (IonE) and active collaborator of the Notre Dame Global Adaptation Initiative. Martina is interdisciplinary researcher, working with scientists across different disciplines. Before joining IonE she was postdoctoral researcher at the Earth Institute of the Columbia University, working with the International Nitrogen Initiative.

Jessica J. Hellmann is the Director of the Institute on the Environment at the University of Minnesota. Hellmann's research focuses on global change ecology and climate adaptation. Before coming to the University of Minnesota in 2015, Hellmann was on the faculty at the University of Notre Dame in the Department of Biological Sciences and served as research director of the Notre Dame Global Adaptation Index.

Stuart Levy is a Senior Lecturer in the School of Education at Federation University Australia. His interests in international relations include the nature and evolution of state sovereignty and the politics of global climate change.

Beth and Stuart have previously co-authored *Climate Change and Order: The End of Prosperity and Democracy*, 2013 and *International Relations: Nurturing Reality*, 2008.

Yudi Li is a graduate student pursuing for M.S. degree in Conservation Sciences under the College of Food, Agriculture, and Natural Resource Sciences at the University of Minnesota. Yudi is also Research Assistant in the Laboratory of Jessica J. Hellmann, focusing on climate change adaptation and ecology.

Hwan-ok Ma is Project Manager at International Tropical Timber Organization, an intergovernmental organization promoting sustainable forest management (SFM) in the tropics. He has over 20-year experience of working with Asia-Pacific regional SFM issues from conservation, restoration, management, and sustainable use with some experience on West Africa and Latin America.

Tek Maraseni is an Associate Professor with the University of Southern Queensland. He has over 25 years of research experience in the areas of value chain analysis, forest governance, and greenhouse gas (GHG) emissions accounting/modelling research in different countries including Nepal, Thailand, Vietnam, Laos, PNG, Fiji, China, India, Bhutan, Myanmar, and Australia.

Mile Mišić is a Ph.D. candidate at the Institute of Political Science at Heidelberg University. His research focuses on energy governance, energy transitions, and comparative environmental politics.

Ross Mittiga is an Assistant Professor in the Instituto de Ciencia Politica at the Pontificia Universidad Catolica de Chile. His research examines the politically disruptive potential of global climate change and the principles, policies, and values bearing on efforts to address it.

Jeremy Moss chaired the UNESCO working group on Climate Ethics and Energy Security and is a recipient of the Eureka Prize for Ethics. His research interests include projects on climate justice, ethical issues associated with renewable energy and climate transitions, including health and climate change. As Co-Director of the Practical Justice Initiative, he leads the *Climate Justice* Research program at UNSW. Moss has published several books including: *Reassessing Egalitarianism, Climate Change and Social Justice*, and *Climate Change and Justice* (Cambridge University Press).

Jane Mummery is a Senior Lecturer in Philosophy at Federation University Australia. Her research examines the ethico-political dimensions of considering more-than-human interests. Her recent books are *Radicalizing Democracy for the Twenty-First Century* (Routledge, 2017), and *Activism and Digital Culture in Australia* (with Debbie Rodan, Rowman & Littlefield, 2018).

Josephine Mummery is a Senior Executive in Australian Government Departments responsible for climate change for 15 years—has led work to coordinate national climate change science investment, build adaptation capacity, including policy and regulation reform, and partner for greenhouse gas emissions reduction. She is currently a University of Canberra Research Fellow, analyzing the climate science–practice interface that enables effective adaptation and risk management policy and decision-making.

Matthew Rimmer is a Professor in Intellectual Property and Innovation Law at the Faculty of Law in the Queensland University of Technology (QUT). He is a leader of the QUT Intellectual Property and Innovation Law research program, and a member of the QUT Digital Media Research Centre (QUT DMRC) the QUT Australian Centre for Health Law Research (QUT ACHLR), and the QUT International Law and Global Governance Research Program. This research was supported by an Australian Research Council Future Fellowship on *Intellectual Property and Climate Change: Inventing Clean Technologies*.

Nicole M. Schmidt is a Ph.D. candidate at the Institute of Political Science at Heidelberg University. Her research addresses comparative EU and global climate change policy, with a focus on adaptation. She has published her research in Global Environmental Change, Review of Policy Research, and Environmental Science and Policy.

Jale Tosun is Professor of Political Science at Heidelberg University. Her research interests comprise comparative policy analysis, public administration, international political economy, and European integration. Her research was published in the *European Journal of Political Research, Journal of European Integration, Journal of European Public Policy, Public Administration*, and *West European Politics*.

Steve Vanderheiden is an Associate Professor of Political Science and Environmental Studies at the University of Colorado at Boulder in the USA. His research focuses upon equity issues in global and domestic climate change policy. He is the author of *Atmospheric Justice: A Political Theory of Climate Change*, Oxford 2008.

List of Figures

List of Tables

1

Introduction to Transformative Climates and Accountable Governance

Beth Edmondson and Stuart Levy

This book examines why accountabilities matter for international climate governance. It brings together diverse approaches to understand how increased accountabilities can improve the effectiveness of climate change governance in promoting orderly social transitions against a backdrop of transforming environments. These are timely discussions because there are real-world consequences of changing ideas regarding how best to achieve effective climate governance. Independent sovereign states are central pillars of the international political system and the practices of the international political economy in which states, intergovernmental organisations and major economic corporations, are dominant actors, can pose institutional impediments to effective

B. Edmondson (✉)
School of Arts, Federation University, Churchill, VIC, Australia
e-mail: beth.edmondson@federation.edu.au

S. Levy
School of Education, Federation University, Churchill, VIC, Australia
e-mail: stuart.levy@federation.edu.au

© The Author(s) 2019
B. Edmondson and S. Levy (eds.), *Transformative Climates and Accountable Governance*, Palgrave Studies in Environmental Transformation, Transition and Accountability, https://doi.org/10.1007/978-3-319-97400-2_1

1

climate mitigation and adaptation strategies. Understanding why and how increased accountabilities can overcome or avoid these limitations, by strengthening mechanisms for climate governance and meeting the popular expectations of informed citizens, may transform the prerogatives and privileges of sovereign states and trigger a further evolution in notions of state sovereignty.

Throughout, authors systematically examine links and tensions between diverse approaches to climate governance, aiming to improve understandings of current and anticipated environmental transformations and to extend capacities to distribute responsibilities for managing their impacts. They pay sustained attention to whether and how understandings and applications of accountability can improve international climate governance mechanisms and institutions. Overall, these chapters hold in mind one of the most pressing questions concerning effective climate change adaptation and mitigation strategies: How does global climate change increase the need for accountable governance?

Effective responses to mitigate the worst consequences of global climate change rely upon global governance mechanisms that commit states to more responsible courses of action. Achieving these will rely upon collective agreements and effective authorities to ensure that internationally agreed targets are met. Orderly and equitable mitigation strategies cannot be achieved through simple broadly based agreements concerning shared responsibilities and common goals of ensuring the longevity of human societies. Such goals necessitate changed international approaches to economic and social policies, including new structures for resolving contested interests (United Nations Development Programme, 2008; Najam, 2005).

This book integrates knowledge and perspectives from biology, environmental and climate science, policy studies, international law, environmental ethics, human geography, economics, philosophy and international relations. Utilising interdisciplinary approaches, they examine new opportunities for developing accountable climate governance mechanisms as climate change-induced environmental transformations disrupt complex natural systems and the global distributions of species, water, arable and habitable land. They also examine their consequences in disrupting current systems of political, economic and social order. They thereby present new interdisciplinary approaches to improving climate governance as environmental transformations and new accountabilities challenge established structures, processes, ideas and values concerning rights, responsibilities and authoritative capacities.

The key aims of this book are to examine the ways that climate change highlights imperatives for increasing the capacities of global governance mechanisms to accelerate the slow rates of progress achieved to date. By creating accountable adaptation and mitigation strategies through new rules, compliance requirements and operational responsibilities, climate governance mechanisms can enhance the effectiveness of the climate change regime complex to achieve better environmental outcomes and improved prospects of sustainable political order. Authors recognise the limitations of existing governance mechanisms and advocate for more systematic and integrated approaches that take heed of ecological and biological systems, human and environmental interfaces, political and economic sectoral interplay and scalable solutions. Throughout, their chapters examine how and why predictable and orderly climate-related transitions depend upon effective and accountable climate governance that incorporates international political leadership.

While slow responses sometimes arise from a lack of political will, it is more often the case that policy-makers become caught in uncertainties concerning risk assessments and fluctuating prospects of effectively managing changed practices (Young, 2002). The political impacts of global climate change go beyond the consequences they pose for the forms, locations and distributions of human societies and their centres of production. They challenge core political values and ideals which seem fitting for the most significant set of issues yet to have faced human civilisations. Problematically, the prospect that existing core political values are challenged by global climate change is a dawning realisation that few political actors readily accept and acknowledge.

This book shows that accountability is an important attribute of effective climate governance mechanisms because it goes to the heart of security, equitable access to resources, cost and burden sharing, and intergenerational protectionary imperatives. In the twenty-first century, accountability is a central feature of many government systems, especially for liberal democratic states, in ways that vastly exceed the earlier historically grounded foundational social contracts between governments and citizens. In the international political system, accountability is not restricted to liberal democratic states as all states seek to ensure that others are accountable to them. These processes of holding to account include recognising the limits of states' territories and jurisdictions, and expectations that states will contribute to orderly international relations by respecting territorial integrity and governmental authority, and ensuring sustainable development practices that do not prevent each other's access to natural resources.

Accountability is an important attribute of effective environmental governance mechanisms because it goes to the heart of security, equitable access to resources, cost and burden sharing, and intergenerational protectionary imperatives. Accountable governments and governance mechanisms imply transparency in terms of being able to see and measure who is accountable, what is being accounted for, and who is held to account and by whom. International climate governance mechanisms enmesh states in networks of mutual accountability. Among the reasons that greenhouse gas emissions targets are debated and international agreements sometimes struggle to achieve extensive levels of implementation are because parties are concerned to clarify the measures by which they might be held to account. Accountability does not, of itself, guarantee that meaningful targets are set or met, and the normative expectations attached to mutual accountabilities can delay or disrupt international agreements. Nonetheless, accountability processes will be essential in orchestrating the many mitigation and adaptation strategies that are required and for advocating for those that are most effective.

As central sources of authority in the international political system, states provide unique sites of accountability for global, regional and local climate change impacts and the governance mechanisms created to deal with them. Climate change governance has also resulted in states accepting new responsibilities for collective mitigation and adaptation strategies—and a contingent range of accountabilities for the goals they set and actions they take to achieve them. Accountabilities are thus perceived as fundamental to states behaving well as they respond to global climate change and collectively establish and implement governance mechanisms, mitigation and adaptation strategies. Although states often fall short of meeting the goals and targets they have set, at least some of these goals and the means of moving towards meeting them would not have been set in the first place without some common ground in accountabilities.

Building enforcement and compliance is a basic problem for states and the international organisations they create to manage, monitor and respond to climate change-driven environmental transformations. The vast majority of international environmental agreements permit only limited enforcement because the broader international context retains the primacy of independent sovereign states. Within international agreements, enforcement and compliance rely upon the positive influence of cooperation. Specific sanctions, such as those that exclude parties from subsequent negotiations, are rare within international environmental agreements. Instead, compliance relies upon states accepting roles as good international citizens, and shaming recalcitrant parties by pointing out instances of poor conduct.

The notion of environmental transformation is especially important for climate change responses that aim to preserve the adaptation capacities of societies, environments and species. It entails shifts in patterns and relationships, including in accountabilities, ecosystems dynamics, the capacities of organisms to adapt, migrate or otherwise respond to environmental or climate changes. Environmental transformation is a key trigger of change in the locations and structures of human and non-human communities within coupled human-environment systems. While it is possible for human societies to make orderly and managed adjustments from one form of organisation and/or production practices as transformation occurs, among non-human communities successful adaptation often requires longer buffering periods. Without periods of transition, natural systems and societies are sometimes dramatically altered with immensely disruptive implications and effects that are intensified by their interwoven proximity.

Case studies examine how issues of scale and interplay affect governance capacities and governmental accountabilities. These case studies engage with real-world and theoretical challenges and are set within a scholarly framework that examines their consequences for climate change governance. Ideas about independent sovereign states and the practices of the international political economy provide key analytic themes.

Central Premises:

1. International governance matters for achieving effective climate governance, mitigation and adaptation strategies, however, it is not currently sufficient to ensure political stability and sustainable societies. As yet, current targets for the reduction of GHG emissions are too low and the governance mechanisms to support these targets are too light or have too limited accountabilities to prevent likely tipping points.
2. International governance relies upon states as central authorities in international policy-making and in making and enforcing domestic laws. Currently, states are also crucial entities holding public accountabilities.
3. Understanding relationships between states, international governance mechanisms, people and corporate actors can shed new light on ways of improving climate governance.
4. Ethics, law and economics are all central elements of effective climate governance mechanisms and the programs they create to manage environmental transformations and their consequences.

There are both direct and indirect environmental transformations arising from global climate change that pose great risks for all societies and all peoples in the twenty-first century. These risks are not evenly distributed across the world. While some people, certain species and specific locations are especially vulnerable to dramatic environmental shifts, all human and non-human species in all parts of the globe will be impacted by climate change and the cascading series of environmental transformations that arise from it. This is not new knowledge, and these observations do not arise from radical new discoveries. Inevitably climate change impacts across the twenty-first century will alter how and where people live, what they eat and how they maintain their economies and societies. These impacts are the results of human decisions and actions that have led to poor governance, low levels of political will and sustained refusals to accept that policy changes are required.

Minimising the extent of climate change impacts through sustained and concerted mitigation and adaptation strategies will require greater levels of accountability in emerging governance frameworks.

For several decades, many of the best minds with the best knowledge from various branches of science, economics, politics and law have urged recognition that climate change impacts pose unprecedented challenges for contemporary societies—and the life-sustaining environmental systems that support them. Their knowledge-based policy networks have argued that interlinked climate change regimes can form regime complexes to streamline policies, projects and governance mechanisms (Climate and Development Knowledge Network, 2015). Establishing shared approaches to climate governance mechanisms can exert stabilising influences as climate change impacts unfold (Loorbach, 2010). Effective climate governance mechanisms can accommodate some of the unpredictabilities and uncertainties that arise from diverse policy sources, climate and environmental changes, adaptation approaches and knowledge paradigms. As the chapters that follow reiterate, a key function of accountable climate governance mechanisms is to support authorities and 'governance systems to deal with uncertainty and surprise' (Pahl-Wostl, Holtz, Kastens, & Knieper, 2010, p. 572).

Accountable climate governance mechanisms increase institutional transparency through collective and individual accountabilities. They can consolidate 'adaptive ...and collaborative management experiences in which the learning and linking functions... of governance are emphasized' (Armitage, Marschke & Plummer, 2008: 87). Accountability is achieved through compliance and monitoring requirements that establish common platforms for states and other stakeholders. Specifically, these can '[build] on customary practices, strategies and institutions' to maximise opportunities for 'capacity-building for learning' (Armitage et al., 2008: 88). Climate governance institutions are expected to balance risks and uncertainties by setting political and other interests in broader contexts of learning and knowledge sharing (Handmer & Dovers 2009; Gallopín, 2006).

In the twenty-first century, effective climate governance mechanisms will contribute to strengthening accountabilities among parties and implementation agents. Climate governance mechanisms will incorporate interdisciplinary knowledge and adopt multiple parallel approaches. They will respond to the complex problems of climate change impacts through multi-scale initiatives that incorporate adaptation and mitigation strategies, for instance, by focussing on concurrent local, regional and international implementation plans and monitoring. As shown in the chapters that follow, effective and accountable climate governance mechanisms will afford equal importance to local, regional and international interests in their mitigation and adaptation efforts.

References

Armitage, D., Marschke, M., & Plummer, R. (2008). Adaptive co-management and the paradox of learning. *Global Environmental Change, 18*, 86–98.

Climate and Development Knowledge Network. (2015). *Annual Report 2015*. www.cdkn.org.

Gallopín, G. C. (2006). Linkages between vulnerability, resilience, and adaptive capacity. *Global Environmental Change, 16*, 293–303.

Handmer, J., & Dovers, S. (2009). A typology of resilience: Rethinking institutions for sustainable development. In E. L. F. Schipper & I. Burton (Eds.), *The Earthscan reader on adaptation to climate change*. New York: Earthscan.

Loorbach, D. (2010). Transition management for sustainable development: A prescriptive, complexity-based governance framework. *Governance: An International Journal of Policy, Administration, and Institutions, 23*(1), 161–183.

Najam, A. (2005). Neither necessary, nor sufficient: Why organizational tinkering will not improve environmental governance. In F. Biermann & S. Bauer (Eds.), *A world environment organization: Solution or threat for effective international environmental governance?* Aldershot: Ashgate.

Pahl-Wostl, C., Holtz, G., Kastens, B., & Knieper, C. (2010). Analyzing complex water governance regimes: The management and transition framework. *Environmental Science & Policy, 13*, 571–581.

United Nations Development Programme. (2008). *Human Development Report 2007/2008 Fighting Climate Change: Human Solidarity in a Divided World*.

Young, O. (2002). *The institutional dimensions of environmental change: Fit, interplay, and scale*. Cambridge: MIT Press.

2

The Limits of States and Changing Regulatory Frameworks—Section One

Beth Edmondson and Stuart Levy

The environmental transformations, transitions and accountabilities associated with global climate change are challenging the structures, processes, ideas and values by which modern societies, economies and sovereign states function. On the one hand, states provide the central organising and authoritative structures of this system and hold central roles in allocating and protecting rights for their peoples and mutual respect for other states. On the other hand, states are unable to meet the challenges of climate change and the needs for effective climate governance through their independent authoritative capacities. In recent decades, states have become central agents in establishing and implementing international organisations, agreements and regimes to

B. Edmondson (✉)
School of Arts, Federation University, Churchill, VIC, Australia
e-mail: beth.edmondson@federation.edu.au

S. Levy
School of Education, Federation University, Churchill, VIC, Australia
e-mail: stuart.levy@federation.edu.au

© The Author(s) 2019
B. Edmondson and S. Levy (eds.), *Transformative Climates and Accountable Governance*, Palgrave Studies in Environmental Transformation, Transition and Accountability, https://doi.org/10.1007/978-3-319-97400-2_2

recognise and respond to global climate change and the environmental transformations that demand adaptation and mitigation strategies.

The chapters that follow examine diverse approaches to understanding, reconciling and implementing climate governance mechanisms and ideas of accountabilities as sovereign states respond to climate change. They examine the significance of these developments for ideas about the rights and responsibilities of contemporary forms of government, relationships between governments and people and the roles of international governance mechanisms. Accountability is examined as a central theme throughout, with attention to its capacity to promote comprehensive responses to climate governance. Overall, these chapters examine the fundamental importance of accountability in ensuring that states behave well as they respond to global climate change and collectively establish and implement governance mechanisms, mitigation and adaptation strategies.

These chapters examine opportunities and approaches to achieve both better environmental outcomes and orderly governance. Fair distributions of costs, responsibilities and accommodations of interests are considered as key factors in effective and sustainable climate governance. These factors are key to avoiding risks of political disorder and widespread conflict as climate-driven environmental transformations occur. However, these chapters also recognise that states—and the climate governance mechanisms they have established to date—have often fallen short of meeting the goals and targets they have set. Extending accountabilities for meeting mitigation and adaptation goals can reduce policy-making delays and decisions-avoidance that have often been evident in international climate negotiations, agreements and implementation approaches.

In Section One, chapters examine some of the key structures and frameworks within which climate governance initiatives have been established. They consider the roles of states and their capacities to be held to account in order to extend understandings of how and why the international climate governance mechanisms established through internationally agreed strategies for responding to the political challenges of climate change are not always effective and outline opportunities for ensuring improved future effectiveness. Their themes include considerations of the structures and frameworks that constrain (and sometimes incapacitate) states as well as the collective governance mechanisms they periodically establish.

We can no longer afford to overlook or ignore the importance of political structures and institutions to achieve effective mitigation and adaptation strategies. The levels at which greenhouse gas emissions targets are set, and whether or not they are achieved, and/or are regarded as tradeable commodities are important to how greenhouse emissions will be reduced, and how global energy production continues (Stern, 2009; Hoffman & Hoffman, 2008). Identifying who will support or limit such initiatives presents additional political issues that also require urgent attention. Climate change policy outcomes will be affected by the status and capacities of those who enter into international agreements and the means by which they follow through in seeking to 'keep their promises' (Young, 2002; O'Neill, 2009).

Edmondson and Levy argue that the Paris Agreement 2015 was mostly a success, especially when considered within the established expectations of the functions and capacities of sovereign states, which privilege their internal interests over external demands. They maintain that the privileged status of sovereignty is used by states to limit the scope of their climate governance accountability. This is both practically and conceptually short-sighted because states hold inherent responsibilities to protect, and this is a key attribute of their sovereign statehood. Consequently, accountability is regarded as the fulcrum for political order as these uncertainties and environmental transformations challenge states.

This chapter considers the implications of environmental transformations driven by climate change that are taking the world into uncharted territories in terms of how societies can be maintained, what production will mean and whether or not sufficient food can be grown to support an unprecedented human population. These are all known problems that can be solved by knowledge of which kinds of governance lead to which kinds of problem-solving. It concludes that a regime complex is a step in the right direction.

Josephine and Jane Mummery's examination of climate governance in the Anthropocene establishes that the relationships between expert knowledge holders and policy makers will need to be differently configured to enable sustainable societies as global climate change unfolds and drives complex environmental transformations. The relationships between people, institutions and nature will also have to shift away from a central focus upon nature as resources and/or environment as the context within which people live. Transdisciplinary knowledge can support new relations, new institutions and new openness at all levels, and this will assist with new solutions through experimentation and new collective accountabilities. Privileging the global environment above domestic environments and production practices and economic interests will then make responses to transboundary problems easier.

Their chapter proposes four imperatives with which to test governance approaches to addressing climate change and apply these to an evaluation of three policy contexts in Australia. From this, they establish the utility and importance of polycentric and decentralised approaches that can compensate for inadequate leadership at the state level and the strong intersectionality and impact of experimentation and anticipation in climate change-oriented governance by sub-state actors.

Matthew Rimmer argues that investor-state disputes highlight the imbalance between states and investors and the privileged status of economics. This imbalance can be resolved by improved transparency, effective law-making and judicial systems, better knowledge of international law and resetting the nexus between domestic and international law. He provides a sustained outline of how civil society groups have exerted influence in pursuit of extended accountabilities in key investor-state trade disputes. The essence of argument here is that citizens, and the groups they form, can make governments and economic actors more accountable and that reforms to trade agreements provide strong opportunities for them to exert accountability-seeking influence.

An underpinning premise of Rimmer's work is that environmental goods are public goods. Within states, the public, that is the citizens, hold particular interests in these goods and their interests are as important as any economic interests held on the basis of property rights. Further, law provides the basis for ascribing priority to the rights of people with regard to public goods and good laws also hold economic actors to account when they seek to breach these rights. He argues that reforms to investor-state dispute settlement clauses in trade agreements are required to ensure they do not enable powerful interests to delay, disrupt and block action in respect of the environment, biodiversity and the climate.

Jale Tosun, Mile Micic and Nicole Schmidt provide insights from within the European Union (EU) about how citizens perceive international climate leadership and delegation of competences for governing climate change within multilevel policy-making systems. This provides an interesting context within which to consider what might limit states' responses to climate change within particular regulatory frameworks. The EU has been able to increase environmental norms by setting EU climate change legislation over member states which then impacts upon the perceptions of state responsibility by citizens.

While the EU has been a leader in establishing international environmental norms, contributing to international environmental regimes and seeking to lead international climate governance, it was found that it is difficult for citizens of the EU to attribute responsibility for climate governance and to know who is accountable. Across the EU, involvement by states in climate policy, and public awareness about this, varies. Significantly, governance is crucial for climate change responses and implementation relies upon public support and so public perceptions of accountability matter. Better understanding how people attribute responsibility and hold policy makers accountable for climate policies, and their preparedness to diffuse competences across different levels of governance, will be necessary to improve environmental outcomes.

References

Hoffman, J., & Hoffman, M. (2008). *Green: Your place in the new energy revolution*. Houndmills: Palgrave Macmillan.

O'Neill, K. (2009). *The environment and international relations*. Port Melbourne: Cambridge University Press.

Stern, N. (2009). *A blueprint for a safer planet: How to manage climate change and create a new era of progress and propserity*. London: The Bodley Head.

Young, O. (2002). *The institutional dimensions of environmental change: Fit, interplay, and scale*. Cambridge: MIT Press.

3

Order and Accountability in Governing Transforming Environments

Beth Edmondson and Stuart Levy

1   Introduction

The 2015 United Nations Climate Change Conference produced the Paris Agreement which has been heralded both as an historic turning point in responding to global climate change and a disappointment (Chan, Brandi, & Bauer, 2016; Kinley, 2017; Suzuki & Hannington, 2017). Known also as COP 21, the 21st meeting of the Conference of Parties who ratified the 1992 UN Framework Convention on Climate Change (UNFCCC), the Paris Agreement brought every state under the same climate change policy framework for the first time. Unlike the 1997 Kyoto Protocol which took eight years to ratify, the Paris

B. Edmondson (✉)
School of Arts, Federation University, Churchill, VIC, Australia
e-mail: beth.edmondson@federation.edu.au

S. Levy
School of Education, Federation University, Churchill, VIC, Australia
e-mail: stuart.levy@federation.edu.au

© The Author(s) 2019
B. Edmondson and S. Levy (eds.), *Transformative Climates and Accountable Governance*, Palgrave Studies in Environmental Transformation, Transition and Accountability, https://doi.org/10.1007/978-3-319-97400-2_3

Agreement came into effect on 4 November 2016 with 195 signatories (https://unfccc.int/process/the-paris-agreement/status-of-ratification). It is notable for shifting climate change debates away from whether or not it is real, or how it might be addressed, and by whom, towards considerations about order and accountability in governing transforming environments. It did not produce a single comprehensive climate change regime, but rather acknowledged the emergence of a regime complex even as analysts identified significant limitations with such structures (Keohane & Victor, 2011; Orsini, Morin, & Young, 2013).

Critics of the Paris Agreement principally point to the modest targets, absence of firm and binding commitments that signatories have to honour and weak compliance mechanisms that are inadequate for limiting the increase in global average temperatures to well below 2°C above pre-industrial levels (Höhne et al., 2017; see also https://climateaction-tracker.org/). Climate change and environmental commentators assert that these deficiencies are likely to result in levels of climate breakdown that will be dangerous to all and lethal to some (Monbiot, 2015; see also Bawden, 2016). These concerns are well founded and will likely become increasingly so as the modelling of global climate change effects is progressively refined and becomes more accurate. Additional critiques also recognise the limits of regime complexes in extending 'strategic interactions between state, business and civil society coalitions' to overcome limited 'domestic mitigation' and improve 'international cooperation' (Ciplet & Roberts, 2017, p. 149).

Accountability processes have emerged as a primary consideration for twenty-first-century climate change governance in response to presumptions in the Paris Agreement that non-state actors will be able to assume increasingly prominent roles alongside states. Imperatives for devising and implementing coordinated mitigation, adaptation and climate resilient infrastructure initiatives have been recognised by a diverse array of actors—and this is itself a significant and positive achievement in the contested political and policy domains of transboundary problems. It has enabled climate change debates to transition beyond the veracity of climate change science and forecasts to mitigation and adaptation possibilities (Chan et al., 2016, p. 238; see also Suzuki & Hannington, 2017). These small steps towards climate change governance mechanisms can support order and accountability as environments transform.

Nonetheless, climate change governance remains difficult and contested because it involves creating frameworks for coordinated activity at the edge of shared knowledge and necessarily invigorates debate about who is responsible for particular initiatives and to whom. This remains the case in all contexts of identifying and assessing indicators of global climate change and in all climate-related policy-making settings.

Climate change governance mechanisms are developed by delegates who gather to discuss what to do about specific emerging problems, how best to address them, and who should assume responsibility. They give rise to challenging questions for those who attend direct discussions and also for those affected by the underlying causes, the emerging effects and the proposed responses and solutions. These challenges persist, irrespective of whether or not resolutions are achieved, because they invariably impact popularly accepted conceptions of progress, prosperity and well-being (Edmondson & Levy, 2013). Discussions concerning how best to respond to global climate change have been occurring for several decades, and in spite of enormous challenges, many agreements have been achieved. However, these agreements have often struggled to be effective and have often not yielded the benefits that were hoped for at their commencement.

Established literature concerning international political interactions between states has observed that changes in environmental conditions create uncertainties that can disrupt progress towards shared goals (Edmondson & Levy, 2013; O'Neill, 2009; Paterson, 2000). Whether occurring in local, regional or international settings, environmental transformations create uncertainties that exacerbate the likelihood of stagnation and decision-making paralysis. Some actors assume increasingly cautious positions when they are uncertain about the possible options, or the potential costs, arising from their individual and shared responsibilities, the responsibilities of others and the points at which these overlap are often intensified by knowledge gaps and capacities dissonance (Sabel & Victor, 2017). Institutional responses then gravitate towards concerns for structural accountability processes that can coordinate and harmoniously orchestrate the activities of diverse actors (Widerberg & Pattberg, 2017). Yet conventionally assumed imperatives for centralised, coordinated and top-down global responses to climate change have often led to complexity-induced institutional stagnation.

2   Climate-Driven Environmental Transformations

Transformations in the natural environment arising from climate changes will themselves impact the capacities of actors to initiate and implement climate change responses in fashions that are likely to be more dramatic than debates about accountability. Climate change-induced environmental transformations are already changing the global distribution of arable land and access to freshwater supplies. While debates about accountabilities may not appear central to the key mitigation and adaptation strategies that are current political and policy priorities, these debates are central to orderly climate change responses. As global climate change and transforming environments impact upon all states and all sectors of societies within them, more diverse actors have sought to engage with debates and to influence policies. The activities of this expanded and diverse host of actors who are motivated to engage in addressing climate change highlight the importance of accountabilities. These actors want to know what can be done and who should be expected to protect them from the most dire of these environmental transformations. In this context, order is a requisite of effective actions, and orderly responses are essential for limiting the risks of major political disruptions that might occur alongside climate change and associated environmental changes.

Scientists are increasingly able to provide greater understandings of both the immediate and long-term climate consequences of current industrial practices and to take account of different developmental trajectories. These give rise to an appreciation that timely responses require transparency and accountability to facilitate knowledge about which initiatives can be most effectively developed and implemented. However, they are compounded by the uncertainties posed by the potentially diverse effects of global climate change, the interests and experiences of diverse actors and the competing logics of action and responses. Refinements in predictive scientific modelling are also making the timelines for effective action on climate change clearer (Höhne et al., 2017). New experimental responses are unlikely to be undertaken in a luxurious universal context of a-political, stable and benign natural environments as the disruptions of climate change and other environmental transformations will pose complex challenges for the orderly development of responses (Martin, Maris & Simberloff, 2016; see also Edmondson & Levy, 2013).

Environmental transformations occur during periods of climatic crisis, species depletion, industrialisation, de-industrialisation or urbanisation and can yield double-edged outcomes. These transformations can be both positive and negative, and most will hold benefits for some species and geophysical systems while others will be adversely affected. The term transformation is often associated with positive changes, especially in relation to changes in human behaviour or other socially constructed categories and entities. As observed by Bauch, Sigdel, Pharaon, and Anand (2016, p. 14560) '[a]s the influence of humans on their environments continues to grow, so too have the resulting impacts of the environment on humans, along with our awareness of those impacts and our efforts to mitigate them'. As demonstrated throughout history, it is possible for societies to make orderly and managed transitions between forms of organisation and/or production practices. However, anthropogenic actions are increasingly destabilising ecological balances and the consequences, beyond destructive weather events, extend to food security, the spread of infectious diseases and economic stability (Galvani, Bauch, Anand, Singer, & Levin, 2016, p. 14502).

3   Adaptation Strategies

Currently developed plans for adaptation strategies are already underway in many parts of the world, for instance the Thames Estuary 2100 Plan in England, the Netherland's programmes of coastal defence and riverine flood abatement and planned settlement relocations in Alaska (Kates, Travis, & Wilbanks, 2012). Such strategies are intended to provide buffering periods of transition to circumvent the most disruptive implications and effects of climate change in specific locations. Nonetheless, to date, barriers to more widespread implementation of adaptation strategies have included uncertainties about the magnitude and risks of climate change, the benefits and costs of adaptation, and the 'institutional and behavioral (sic) barriers that tend to maintain existing resources systems and policies' (Kates et al., 2012, p. 7158). Addressing transforming environments requires both mitigation and adaptation strategies that recognise complexities in natural systems, the relational nexus between human and non-human species, and social, economic and political interlinkages. An undesirable outcome of relying upon adaptive strategies might see societies 'doomed to linger perpetually in the neighbourhood' of a critical climatic transformation or tipping point (Bauch et al., 2016, p. 14562). Notably, except for major natural disasters, environments transform most rapidly when human practices and interventions deplete forests, re-route waterways and encroach upon coastlines impacted by sea-level rises, or changed rainfall patterns subsequently create new deserts or trigger landslides.

Environmental transformation entails shifts in patterns and relationships, ecosystem dynamics and the capacities of organisms to adapt, migrate or otherwise respond to environmental or climatic changes, and efforts to understand these draw upon a broad array of perspectives and approaches. The notion of transformation is especially important for climate change responses that aim to preserve the adaptive capacities of societies, environments and species. Recent reported studies have found that 'if nations do no more than they have pledged so far to reduce their greenhouse gas emissions—and warming consequently shoots past 3 degrees by the end of this century—6 percent of all vertebrates would be at risk. So would 44 percent of plants and … 49 percent of insects' (Cushman Jr. & Banerjee, 2018). The consequences of species extinction on such a scale would be profoundly transformative to agricultural practices and food chains. Global impacts however are experienced most directly in regional and local contexts, and it makes more sense to talk about multiple environmental transformations than to homogenise on a global scale. Doing so might help to untangle some of the complexities of responding to global climate changes that are causing/leading to locally and unevenly experienced environmental transformations.

Although not all environmental transformations are caused by climate change and direct human intervention, activities such as forest clearing or river damming are significant factors. Climate change is a leading cause of environmental transformations and a rise in global temperature of 2°C above pre-industrial levels means that:

> large parts of the world's surface will become less habitable. The people of these regions are likely to face wilder extremes: worse droughts in some places, worse floods in others, greater storms and, potentially, grave impacts on food supply. Islands and coastal districts in many parts of the world are in danger of disappearing beneath the waves. A combination of acidifying seas, coral death and Arctic melting means that entire marine food chains could collapse. On land, rainforests may retreat, rivers fail and deserts spread. Mass extinction is likely to be the hallmark of our era. (Monbiot, 2015)

The forms of industry, patterns of urbanisation and societies based upon resource intensive consumption are leading direct causes of environmental transformation and also contributing causes of climate change. Their impacts are direct in terms of overtaking agricultural lands or demanding that rivers be dammed, diverted or drained to support the need for water. These are complex realities, and as observed by Brand and Gorg (2013, p. 110), '[g]lobal environmental change and related politics are increasingly seen as taking place in a complex field in which several ecological processes are interlinked – e.g. climate change, biodiversity, water, and land-use change – and these processes are deeply interconnected with societal processes, such as food supply and nutrition' alongside economic and financial factors.

4 Emerging Frameworks for Global Responses to Climate Change

At Paris, as before, aspirations for creating a comprehensive and overarching global framework for climate change management proved elusive as the political complexities of seeking to regulate and operationalise rules-based responses challenged consensus among states (Dimitrov, 2016). These complexities were not unique to the Paris

Conference and persist because global climate change can be best understood as an interlinked series of wicked problems in environmental change and policy settings. Wicked problems are understood to be particularly resistant to resolution because of their complexity and evolving appreciation. As noted by Hulme (2009, p. 333 cited in Keohane, 2015, p. 20), 'framing climate change as a "mega-problem" almost inevitably "led us down the wrong road" in earlier efforts to create an effective climate change regime at international negotiations and established close binding cycles of political "log-jam[s] of gigantic proportions"'. In such contexts of 'high uncertainty and policy flux', regime complexes can provide effective initiatives to address and support new solutions to persistent problems (Keohane & Victor, 2010, p. 2). These characteristics are especially valuable when 'the first best possibility – a coherent, effective, and legitimate comprehensive regime, with sufficient flexibility to accommodate needed change in response to new situations and information – is politically unattainable' (Keohane & Victor, 2010, p. 23). Consequently, the Paris Agreement represents something of a compromise. The pursuit of a comprehensive top-down regime that had defied effective articulation for decades was relinquished for a more loosely defined regime complex to scaffold bottom-up initiatives from a broad array of actors.

The 2015 Paris Agreement acknowledged the global climate change complex had 'emerged as a result of many state choices at different times and on different specific issues', rather than having been systematically and 'comprehensively designed' (Keohane & Victor, 2010, p. 2). While a single, comprehensive, integrated regime might offer advantages in terms of coordinated responses and governance structures and lead to economies in negotiating the terms and targets of many separate regimes, the current regime complex is not without opportunities to extend accountabilities. The efficacy of the current climate change regime complex rests upon harnessing overlapping regimes to encompass the varied aspirations of diverse actors and support compliance mechanisms and standards that transcend the potentially self-interested regulatory undertakings of individual sovereign states. In these ways, the climate change regime complex can move beyond 'merely rationaliz[ing] an already emerging system of domestically driven climate policy' (Falkner, 2016, p. 1119). These challenges are compounded by '[g]lobal governance architectures, legal and institutional, [that] are said to be fragmenting' (Ruggie, 2014, p. 5). Amid the excitement of the 2015 Paris Agreement, lingered concerns that an 'already weak system of global governance' is 'becoming more so' (Ruggie, 2014, p. 5; see also Biermann, Pattberg, & van Asselt, 2009). This observation resonates with the often-lamented assessment that the Paris Agreement lacks mandated targets supported with compliance and enforcement mechanisms and instead relies upon voluntary commitments that are evaluated via periodic consultative, non-punitive, facilitative self-reporting (Karlsson-Vinkhuyzen, Groff, Tamás, Dahl, Harder, & Hassall, 2017).

Regimes involve strong and deliberate accommodations of diverse interests, which reinforce their institutional and normative roles. In the case of climate change, regimes are comprised largely of international institutions and programmes and the broader international agreements upon which they are based. Regimes create and maintain both formal and informal rules and procedures which are expressed in their routine undertakings and directly influence the scope, breadth and extent of their programmes and regulatory settings. In both formal and informal ways, regimes regulate and shape decisions and structures, provide foundations for new decisions and policies, and provide conduits for shared knowledge.

To succeed, the 'normative and structural components of a regime' must align with other 'major developments in the international community' to preserve their abilities to resolve problems in relationships between parties and to set new goals or adopt new approaches in response to new knowledge (Stokke & Vidas, 1996, p. 24). The success of regimes depends upon their abilities to increase cooperation—which establishes important pre-conditions for accountability. Understood on these terms, the basis for universal acceptance of the Paris Agreement was its move away from the mandated mitigation targets approach of the Kyoto Protocol in favour of voluntary nationally determined commitments (NDCs) supported by a range of civil society actors to whom states would be domestically responsive and accountable.

As a loose coupling of specific regimes, a regime complex provides opportunities for many different political, social and economic actors to pursue shared strategies and to collaborate towards more effective regulatory frameworks (Keohane & Victor, 2010). In this way, a regime complex can interlink established regimes and bring together multiple initiatives to enable collaborative policy input from an expanded range of actors that do not rely primarily or solely upon states. In this regard, regime complexes can create additional political challenges as overlapping interactions among their constituent parts can create or exacerbate perceived competing interests among stakeholders (Orsini et al., 2013, p. 31; see also Ciplet & Roberts, 2017). These overlaps emerge from the complex interdependencies and increasing density of interactions that occur within the maturing international system as states, intergovernmental and non-governmental organisations no longer act in isolation from each other.

A regime complex can support coordinated mitigation and adaptation efforts by enabling states to continue to provide the underpinning sources of authority upon which multilateral agreements are established and maintained. As states are uniquely placed as agents for accountability in the current international political system, acknowledging their privileged status as central sources of authority might still prove beneficial for effective climate change responses. The governance mechanisms they have created have already resulted in a diversity of actors accepting new responsibilities and accountabilities across a wide range of issue areas, such as human rights and international trade, and include greenhouse gas emission reductions to address climate change. A regime complex extends these dynamics to encourage institutional flexibility through both covenant and contract aspects of governance. In situations such as the international political community currently faces in tackling climate change challenges, regime complexes can help to reconcile tensions arising from expectations gaps and uncertainties in knowledge or possible consequences.

Efforts to improve the effectiveness of global governance mechanisms have sought to expand compliance and regulatory frameworks (Ruggie, 2014; Sabel & Victor, 2017), and as states pursue a range of approaches in adapting to specific climate change impacts, similar experimental efforts can be expected. In acknowledging, and relying upon, voluntary contributions towards addressing global climate change by state and non-state actors within a regime complex, the principal question that emerges concerns how order and accountability are to be ensured. This is particularly true with further acknowledgement that the natural environment is in the process of climate change-induced transformations. The long-term uncertainties these produce 'makes it difficult for government to assess where their national interests lie' and will impact the activities of actors within the regime complex (Falkner, 2016, p. 1110). As observed by Sabel and Victor (2017, p. 16) '[w]here uncertainty is high, and actors, unsure of what outcomes are possible, are unable to specify reliably their own interest nor understand with precision the interests of others, experimentation and learning are better means of advancing'. For these reasons, it is anticipated that a regime complex will encourage and facilitate creative bottom-up responses from a broad array of engaged actors to expand accountable and scalable solutions.

As shown in Fig. 1, regimes work to broaden the base of support for governance mechanisms and regulatory frameworks and are generally characterised as 'international institutions consisting of agreed-upon principles, norms, rules, procedures and programmes that govern the interactions of actors in specific issue-areas' (Wettestad, 1999, p. 7). They contribute to new norms of behaviour among states and within international institutions through their capacities to promote collective patience and persistence. By working with states and recognising their pre-eminent political status, regimes can set new expectations that states should better utilise their jurisdictional capacities to achieve improved environmental outcomes.

Fig. 1 A simple representation of how regimes support accountability

Knowledge- and expert-based communities are important contributors and actors in climate change regimes and create sophisticated policy networks that help to shape and guide the institutions that have been established to support mitigation and adaptation approaches. They build participants' and institutional knowledge, identify and respond to new problems, modify targets and project into the future. These contributions are especially important in overcoming the policy stalemates and obstacles in negotiations that arise from uncertainties and for identifying and validating best practice climate change responses. Embedding and threading climate change and governance expertise throughout the policy-making structure of regimes (see Fig. 1) optimises informed and effective outcomes and accountability processes. In the post-Paris Agreement regime complex, an important contribution of epistemic communities will be to inform, facilitate and support the technical examination processes (TEPs) established to examine mitigation and adaptation initiatives. At technical expert meetings (TEMs) 'State and non-State stakeholders share policies, practices and technologies and address … finance, technology and capacity building' (Chan et al., 2016, p. 240). In an oblique acknowledgement of the emissions gap, the UNFCCC notes that '[t]he technical examination process consists of regular in-session thematic technical expert meetings and focused follow up work to be conducted by Parties, international organizations and partnerships … in order to implement scalable best practice policies and bridge the ambition gap' (http://unfccc.int/resource/climateaction2020/tep/index.html).

These networks of policy-makers and knowledge holders provide expertise and exchanges of ideas through shared knowledge. By facilitating learning across their networks, they can support effective climate change adaptation and mitigation strategies and promote new accountabilities for their implementation (Edmondson & Levy, 2013). Within a regime complex, institutions and agencies can both recognise that climate change and associated transformations of environments and societies pose persistent governance problems and become accountable for leading problem-solving approaches.

Knowledge-based collaborative policy networks play particular roles in maintaining or re-activating the decision-making momentum that can be lost or eroded as different issues demand attention. Through collaboration, policy networks provide new opportunities for developing shared understandings across diverse fields of inquiry and across different governance scales. By making decisions and developing strategies, these networks contribute to the formation of governance mechanisms. The opportunities they create for structured dialogue between parties create formal and informal shared accountabilities and enable the emergence of common interests (Edmondson, 2015). By preserving the breadth of negotiations, maintaining goodwill among parties, supporting planning processes and establishing implementation strategies that are supported by stable shared norms, international negotiations and governance mechanisms can progress towards more accountable and effective climate change governance (Edmondson, 2015).

Following the Paris Agreement, a regime complex comprised of interlocked climate change regimes has been accepted as the most likely and politically acceptable approach by which international order might be sustained while simultaneously supporting and enabling effective collective responses to current and predicted climate change impacts that threaten to disrupt familiar forms of political authority and human association. Extending knowledge about these challenges and the capacities for interlocked regimes to provide effective climate change governance mechanisms can contribute to the efficacy of the post-Paris climate change regime complex. The central characteristics and advantages of interlocked regimes can be identified as:

- multiple linked agreements;
- collectively maintained authorities;
- overlapping decision-making structures;
- shared and overlapping targets, problems and/or policy responses;
- shared multilayered knowledge bases; and
- responsiveness to the 'veil of uncertainty' and/or 'shadow of the future'.

These characteristics underscore the advantages that interlocked regime networks confer to building effective and accountable global governance mechanisms. Beginning with assessments of the potential benefits of interlocked institutional mechanisms across key international climate change regimes provides a valuable access point for analysing the roles of policy networks within global climate change regimes.

Such analysis can also illuminate opportunities to minimise costs across multiple international agreements and streamline their implementation and review processes. Doing so is likely to support orderly and predictable conduct in managing the inter-related problems that arise during the development of mechanisms to support global governance during environmental and climatic transformations.

Following the Paris Conference, it is readily apparent that effective climate change governance requires internationally articulated, environmentally responsible and accountable regulatory frameworks underpinned by the authority of sovereign states in a form of 'soft-reciprocity' (Falkner, 2016, p. 1124) encouraged, supplemented and motivated by non-state civil society actors (Karlsson-Vinkhyzen et al., 2017; Kinley, 2017). What remains to be constructed, post-Paris Agreement, is the expanded 'super-structure' of climate change governance that can support monitoring, reporting and review processes and routines that create predictable communications and collaborative policy developments between state and non-state actors. These can reinforce state and non-state actor relationships that promote effective problem-solving (Falkner, 2016; Karlsson-Vinkhyzen et al., 2017; Kinley, 2017). These processes and organisations can increase acceptance of new responsibilities and accountabilities as states set goals and take collective actions with state and non-state partners.

5   Accountability and Orderly Responses

Against a backdrop of transforming environments, climate governance mechanisms can become more effective through increased shared accountabilities. This is critical following the Paris Agreement as states submit nationally determined commitments (NDCs) that are respectful of diverse domestic circumstances, but are also responsive to global governance imperatives to contain climate change. The practical consequences of changing ideas about political authority and accountability, including the scope and sources of rules for regulating behaviour, currently limit effective climate change governance. Overcoming these impediments rests upon understanding why and how increased accountability might transform notions of sovereign state authority to sustain orderly responses to global climate change. Among states, the processes of holding to account include recognising the limits of territory, contributing to orderly international relations, respect for territorial integrity and governmental authority, and ensuring sustainable development practices that do not inhibit or prevent each other's access to natural resources in contexts of environmental transformations.

Accountability is an important attribute of effective environmental governance mechanisms because it goes to the heart of security, equitable access to resources, cost and burden sharing, and intergenerational protectionary imperatives. Narrowly, accountability can be understood as formal mechanisms to ensure states and actors comply with commitments prescribed in international laws and treaties. More broadly, it entails providing explanations of conduct supported by mechanisms that can range 'from educative to punitive' (Karlsson-Vinkhuyzen et al., 2017, p. 2). Within contemporary liberal democratic states, accountability exceeds the historically grounded social contracts that founded relationships between governments and citizens. Through international climate change regime complexes, climate governance accountabilities extend peer accountabilities to all political actors.

Climate change mitigation initiatives have long been understood as requiring fundamental changes to the conduct of modern societies in their interactions with the natural environment and relationships with natural resources (Martin et al., 2016; see also Eckersley, 2004, 2005). These initiatives impose immediate costs to produce anticipated intergenerational benefits of which there are no immediate and obvious indicators and have produced a 'malign politics of too little action' (Keohane, 2015, p. 20) by governments, corporations and the publics to which they are accountable. Further, there are clearly understood free-rider incentives in a global economy to delay adopting costly mitigation initiatives (Falkner, 2016) and devising sanctions to encourage mitigation efforts are necessarily divisive, difficult to enforce and reliant on soft measures that impose reputational costs (Widerberg & Pattberg, 2017).

Climate change adaptation responses can be local, immediate and obvious in the production of a public good and as such are likely to be most widely promoted by civil society actors and enthusiastically endorsed by actors with clear accountabilities such as governments to electorates and corporations to shareholders. While adaptation is a critical aspect of climate change responses that help to sustain community and social aspirations of prosperity and well-being, they are nonetheless reactionary and do not address the underlying causes of climate transformations. Focusing primarily upon initiatives that address adaptation imperatives locks communities into escalating cycles of change and costs associated with responding to climate transformations.

Climate change responses that focus upon the development of climate resilient infrastructure are expensive, uncertain and beyond the reach of many actors outside the developed world. Infrastructure development can improve energy security and reduce the production of greenhouse gases. It can also generate universal public goods that require protecting from free-rider abuses to achieve sustainable economic production and prosperity arising from clean technology industries. The potential to re-create, or create new, relationships of dependency that ultimately jeopardise order is a plausible consequence as 'only six countries account for about 85% of all R&D investment' (Keohane & Victor, 2010, p. 23).

How climate change challenges are understood underpins the strategies and initiatives of actors within the regime complex and the regulatory mechanisms that are established. Mitigation, adaption and infrastructure development are, by themselves, divisive among states and non-state actors, and how these are framed within the regime complex will be important for enabling progressive and effectively balanced responses. As noted by Dimitrov (2016, p. 3) '[d]eveloped countries were united in Paris in seeking a global agreement that focuses on mitigation; skirts adaptation (important to least developed and island states); preempts [sic] legal obligations on finance, compensation, and technology transfer; and includes strong international transparency for national mitigation actions'. These observations sit comfortably alongside earlier realisations that environmental governance mechanisms that are more concerned with achieving and implementing mitigation strategies tend to privilege resilience as an indicator of environmental and governmental transitions capabilities.

International climate change governance and efforts to regulate behaviour in international settings are achieved through norms, customary international law and common principles that guide the actions and policy decisions of contemporary states. Effective international climate change governance regulates behaviour through international norms and customary international law to establish common principles that guide the actions and policy decisions of governments and international organisations. These norms and regulatory approaches shape the parameters of actions pursued by a diverse array of governments, international organisations, economic and civil society actors. The frameworks they create and the rounds of negotiation they support can, over time, expand cooperative goals setting and support readily scalable multilateral climate governance initiatives. In so doing, international climate change governance mechanisms can position themselves as key sources of accountability for establishing and implementing international mitigation and adaptation approaches (Chan et al., 2015). These factors are especially the case for governance mechanisms that have been endorsed by the UNFCCC, whose legitimacy and authority is underpinned by universal state membership.

Central to accountability mechanisms across a regime complex is the hope that coordinated responses to transnational, cross-jurisdictional issues will be achieved and the risks of political disorder, inefficiencies, increased transaction costs and possible conflict will be reduced (Chan et al., 2016). Accountabilities are thus perceived as fundamental to the behaviours of state and non-state actors alike as they respond to global climate change and collectively establish and implement governance mechanisms and strategies. States—and the international mechanisms, agreements, treaties and protocols that were established to address climate change, such as reducing carbon emissions—have often fallen short of meeting the goals and targets they set. Nevertheless, at least some of these goals and the means of moving towards their attainment would not have been identified in the first place without the common ground of mutual accountabilities.

Achieving the lauded successes of the Paris Agreement and the commitment of states relied upon replacing legally enforceable mitigation targets with communally crafted compliance mechanisms that are 'non-judicial, non-confrontational and consultative' (Karlsson-Vinkhyzen et al., 2017, p. 1). Its success in achieving near universal acceptance and rapid ratification by states was facilitated by the adoption of Nationally Determined Contributions (NDCs) through which states set their own climate mitigation targets rather than being accountable to centrally mandated mitigation targets (Falkner, 2016). This change resolved the long-standing deadlock arising from the distributional challenge of allocating levels of responsibility for addressing climate change. The NDCs mechanism of the Paris Agreement privileges the sovereignty of states by firmly locating climate change actions within nationally self-determined commitments. While states are legally obligated to submit NDCs, and report their actual achievements every five years, meeting these voluntary commitments is not a legal obligation (Falkner, 2016, p. 1118). States' accountabilities to meet their NDCs then relies upon maintaining a communal sense of urgency in addressing climate change through softer forms of suasion within the climate change regime complex.

It is optimistically presumed that 'naming and shaming' recalcitrant states by peers and civil society actors will encourage a 'norm cascade' of positive and effective initiatives even though 'effective mutual accountability through peer-pressure among states may seem rather slim' (Karlsson-Vinkhyzen et al., 2017, p. 3). This strategy rests upon three assumptions. The first is that the actors who compose civil society are themselves unified with commonly aligned or sympathetic agendas and are less fractious that the community of states. The second is that the currently reported groundswell of support for action on climate change (Chan et al., 2015, 2016; Suzuki & Hannington, 2017; Widerberg & Pattberg, 2017) by civil society and electorates whose interests can be notoriously fickle and divisive can be extended beyond developed states and sustained across many decades. The third is that all states will become increasingly responsive to national and transnational civil society expectations and demands that may not align with popularly conceived national political and economic interests. This second assumption poses a latent challenge to states' accepted sovereign rights to non-intervention or interference in domestic affairs and may herald a necessary evolution in state sovereignty (Edmondson & Levy, 2013) in further recognition that 'all must participate in solving the huge global challenge posed by climate change' (Kinley, 2017, p. 13).

Implementing accountability mechanisms in the Paris Agreement rests with the NDCs of states as they articulate individually determined and voluntary incremental commitments to reducing greenhouse gas emissions and meeting the climate goals of the UNFCCC. The adoption of NDCs has been considered an improvement on the top-down pledge-and-review provision of the preceding Kyoto Protocol which all too often became pledge-and-revise as states failed to meet their targets and withdrew from the Protocol (Keohane, 2015). The NDCs are intended to commit states to incremental action so that even if they fall short of their specified targets the incremental nature of the NDCs should have a ratcheting effect on their actual achievements (Falkner, 2016; Höhne et al., 2017). However, the NDCs are not legally binding and could become pledge-and-ignore commitments if the optimistically presumed pressure from civil society actors does not sustain states' adherence across many future decades.

Developing accountability mechanisms that compel states, without alienating them, to meet and incrementally raise the levels of their NDCs while still accommodating prevalent conceptions of state sovereignty is critically important. It is the possession of sovereign authority that makes states, among all the actors within the climate change regime complex, ultimately accountable to each other and their citizens for ensuring their individual and collective longevity. This is an immensely powerful dynamic that can and should be utilised in progressing accountable climate governance and for addressing the widely acknowledged emission gap (https://climateactiontracker.org/) that exists between the first round of NDCs and those required to meet the Paris Agreement target of less than 2°C. As observed by Kinley (2017, p. 13) the initial NDCs submitted by states as part of the Paris Agreement 'are inadequate to the task. They are projected to get the world somewhere near 3°C of warming'. This is far short of the aspirational 1.5°C and will have far-reaching and potentially devastating transformative consequences for the natural environment, animal life and human societies.

While states are acknowledged as having been 'reluctant to build strong accountability regimes at the global level' there remain sound reasons to believe that peer review mechanisms could effectively augment states' accountabilities (Karlsson-Vinkhuyzen et al., 2017, p. 1). Identifiable sources of increased internal accountabilities are evident in states' internal responsibilities for maintaining informed publics, orderly domestic institutions and transparent internal government audits. A collaborative path, respectful of state sovereignty, could well be charted if assessments are correct and the initial NDCs submitted by states under-report the contributions made by civil society which will then make it easier for states to meet and exceed their commitments (Höhne et al., 2017, p. 25). Although in terms of addressing the emission gap, as observed by Chan, Brandi and Bauer (2016, p. 246) '[a] major challenge to estimating the contributions by non-State actions is a lack of clear targets and standards for monitoring, reporting and verification'.

A further consequence of transforming the nature of distributional mitigation targets into the NDC model has been the culmination of a long-term trend towards the entrenchment of neoliberal attitudes towards the environment and justice (Ciplet & Roberts, 2017). Recognising the desirability of sustainable development, safe-guarding states' access to resources, extending the roles of private and civil society actors towards future-oriented global responsibilities and away from past-oriented historical culpabilities can all be considered positive achievements. Less evident in international agreements has been attention to issues of distributional impacts as states have focused primarily upon their own interests. In moving states beyond deadlock, the Paris Agreement 2015 did not pay particular attention to distributional justice and equity or the precautionary principle.

6   Order in a Transforming Natural Environment

Orderly governance supported by effective political mechanisms will impact the rights and capacities of states and change their economic activities and structures as environmental transformations arise from climate change consequences. Within the post-Paris Agreement climate change regime complex, states retain their rights to claim exclusive and privileged use of the natural resources within their territories; however, there remains scope for dispute among states regarding these resources claims. Further, states retain their capacities to claim they authoritatively represent their citizens' interests; however, this also may be disputed. One such example followed the announcement in 2017 that the USA intended to leave the Paris Agreement. A coalition of more than 1200 American businesses, cities, states, universities, indigenous groups and other signatories reacted by affirming their commitment to the Paris Agreement (www.wearestillin.com).

In climate change governance, the imperatives for accelerating progress in adaptation and mitigation strategies and the development of new low-carbon infrastructure support new considerations of institutional accountability. Accountable climate governance mechanisms support collaborative policy networking and establish structures that enable targeted policy implementation and monitoring as their networks create new trust-building mechanisms. By creating and paying attention to the rules, compliance requirements and the operational responsibilities of established mechanisms, climate governance institutions improve long- and intermediate-term climate change management. As observed by Brand and Gorg (2013, p. 117), these international institutions can serve 'as regulating authorities, as expressions and stabilizers of global power relations, and as political terrains to deal with conflicts'. Accountable climate change institutions may also increase individual and collective responsibilities for developing and implementing mitigation and adaptation strategies. Doing so requires organisational flexibility within climate governance mechanisms and collective recognition of their authoritative status (Eckersley, 2005; Edmondson, 2015; Nelson, 2009; Young 2002). Accountable institutions can achieve a good 'fit' between targets across multiple scales and can also support appropriately located learning-based compliance monitoring. Integrating new and expanded expectations of accountability into international climate change mechanisms can achieve rapid implementation of local, regional and international mitigation and adaptation projects and the maintenance of long-term collaborative policy networks.

Accountable climate change institutions rely upon systematic and integrated approaches to recognise the diverse needs of humans and non-humans, their relationships and natural-social systems nexus complexities. Accountable climate governance institutions can also use more direct forms of monitoring to move beyond risk avoidance approaches to emphasise the benefits of positive change and extend their adaptive capacities (Edmondson & Levy, 2013). The development of integrated responses that accommodate ecological and biological systems alongside human societies requires mechanisms to evaluate the effectiveness and accountability of solutions proposed by a diverse range of actors.

Innovative ways of sharing governmental authority will be essential for progressing the collective accountabilities required for effective climate change governance. In this way, accountability provides a central notion and ethical lens through which to examine real-world and theoretical accountabilities and engage with different ways of understanding accountable relationships. Ideas about the accountability of governance mechanisms then provide frameworks for discussions about the roles and relationships between different entities. These approaches ensure sustained attention to how and why accountability implies transparency—being able to see and measure who and what is being accounted for, and who is held to account and by whom.

It is currently impossible to predict when or whether targets will be met, or whether chosen governance methods are optimal. These ongoing uncertainties highlight the need to know more about how to ensure compliance and how to achieve equitable distributions of benefits, burdens, accountabilities and leadership in pursuing new initiatives. They also highlight the importance of extending knowledge of how to create international institutions with flexibility, durability and monitoring/compliance mechanisms to achieve equitable distributions of benefits, responsibilities and burdens in managing the impacts of global climate change. Establishing new models for interlocked regimes should make it possible to develop and implement effective and durable international policies and governance mechanisms, including in currently problematic areas of mitigation and adaptation to progress global climate governance beyond the current loosely coupled regime complex. Doing so requires new examinations of the characteristics of institutional effectiveness; relative advantages of soft and hard law; monitoring and compliance mechanisms; knowledge networks; epistemic communities; policy networks; stages and processes of negotiations, bargaining and cooperation.

Beyond evaluating specific initiatives, and identifying which audiences might hold which new initiatives to account, are questions about how the current regime complex avoids creating fragmented and uncoordinated responses. Effective and coordinated global responses to climate change requires considerable effort in three fields: mitigation responses to alleviate the underlying causes of global warming and climate transformation; adaptation responses to permit communities to be sustained within transforming climates; and the development of climate resilient infrastructure that perpetuates aspirations of progress, prosperity and well-being. With initiatives in these fields emerging from diverse actors and sources of interests across the regime complex, coordinated responses require accountability mechanisms that ensure actors engage with all three fields and that none are neglected to prevent imbalances that compromise global responses. Prospects of imbalance arise from conventional framing of these fields and the realities of changing the habitual practices of billions of people. Clearly developed accountability mechanisms are needed to promote a well-balanced suite of responses.

7   Conclusion—The End of the Beginning

Accountability is a central aspect of the efficacy of regime complexes because it influences their coherence, effectiveness and epistemic qualities (Keohane & Victor, 2010). In effective regime complexes, accountability processes interlock the activities of constituent regimes and their respective parties so that negative consequences such as 'gridlock rather than innovation', 'critical veto points' and 'forum-shopping' can be avoided (Keohane & Victor, 2010, p. 17). Consequently, clarity in the development of accountability processes and mechanisms in the climate change regime complex are recognised as essential for promoting the most promising climate change responses and avoiding fragmentation.

Mitigating the worst consequences of global climate change relies upon global governance mechanisms that commit states to more responsible courses of collective and individual action. Such mechanisms rely upon collective agreements and effective institutional authorities. Broadly based agreements concerning shared responsibilities and common goals will not be sufficient for orderly and equitable climate change mitigation strategies.

Enforcement and compliance are also basic problems for international organisations as they seek to manage, regulate and monitor global climate change (Hempel, 1996). Within climate change agreements, compliance relies upon parties accepting their obligations as specific sanctions are rare.

Accountability will be central in post-Paris climate governance developments to underpin the convergence of norms, policies and practices, from a diversity of actors, to create a comprehensive and integrated suite of effective climate change responses. Negotiations and agreements to improve climate change responses often proceed slowly, sometimes because of a lack of political will, but more often because policymakers become caught in uncertainties concerning likely risks, the merits of different risk assessment approaches or possible opportunities to develop and pursue changed practices (Young, 2002). How levels of greenhouse gas emissions targets are set, whether or not they are achieved, whether and how they are regarded as tradeable commodities all matter for how greenhouse emissions will be reduced, and also for which forms of global energy production continues, in which locations, and for which uses (Hoffman & Hoffman, 2008; Stern, 2009). The status and capacities of the actors who enter into and support international agreements and the ways they seek to 'keep their promises' affect the levels of emissions reductions that are achieved and the structures and accountabilities of global climate change mechanisms.

Over the past decades, much has been written and speculated about the consequences of climate transformations for developed and developing societies and the habitability of the planet. Alongside these an identified 'groundswell' of support for action to address climate change has emerged (Suzuki & Hannington, 2017; see also Chan et al., 2015; Falkner, 2016). Considerable scientific work has been sponsored by the UNFCCC, and its complexity has limited broad dissemination and produced keen debate. The outcome of these parallel activities has been a growing civil society appreciation of the impending consequences of global climate change. Depictions of fictional dystopian futures have also contributed to raising broad-based awareness about potential future disorderly impacts of climatic and environmental transformations. They too have played roles in galvanising the observed groundswell in climate action. As noted by Keohane and Victor (2010, p. 3), '[s]tates construct international regimes on the basis of their interests' which 'reflect the interests of the major constituencies' shaped in turn by prevailing information and beliefs 'that exert influence over state leaders'. Similarly, global business and commerce are increasingly prepared to act on climate change and are becoming advocates of action in response to climate science rather than sources of cynical resistance despite ongoing public reservations about the pervasiveness of green-washing.

States, businesses and civil society appear now prepared to address climate change in concert. This unifies the realms of politics, economy and society. The emergence, post-Paris, of an acknowledged and recognisable regime complex that provides an early framework for harnessing and coordinating the activities of these diverse actors signifies a transformation in global climate change governance (Chan et al., 2016). Paradoxically, the uncertainties associated with transforming natural environments could prove to be a boon. The unpredictable effects and consequences of a climate in transition might keep efforts to address climate change at the forefront of public consciousness, strengthen the relevance of penalty defaults and maintain civil society interest in transparent accountability processes to identify best practice solutions. It can be reliably anticipated that challenges to order will arise from the effects of transforming environments. These effects will likely lead some actors to prioritise their activities in visible adaptation rather than less visible mitigation.

Optimistically, acknowledging the current regime complex as a framework by which to orchestrate global climate responses could be likened to the 'end of the beginning' insofar as success is not assured and much remains to be done to achieve the Paris target of limiting the increase in global average temperatures to below 2 °C above pre-industrial levels. Doing so requires fresh thinking and fresh practices that inform and support each other (Ruggie, 2014, p. 15) to achieve more transparent, orderly and accountable approaches and outcomes. These processes require bottom-up and scalable creativity alongside thoughtful re-examination of past proposals that might be adapted and operationalised. These goals can be achieved by diverse actors working alongside states to evolve their focus on sovereign authority to support their ongoing primacy as global guardians (Edmondson & Levy, 2013).

References

Bauch, C. T., Sigdel, R., Pharaon, J., & Anand, M. (2016, December 20). Early warning signals of regime shifts in coupled human-environment systems. *Proceedings of the National Academy of Sciences of the United States of America, 113*(51), 14560–67.

Bawden, T. (2016, 8 January). COP21: Paris deal far too weak to prevent devastating climate change, academics warn. *Independent*. https://www.independent.co.uk/environment/climate-change/cop21-paris-deal-far-too-weak-to-prevent-devastating-climate-change-academics-warn-a6803096.html. Accessed May 1, 2018.

Biermann, F., Pattberg, P., & van Asselt, H. (2009, November). The fragmentation of global governance architectures: A framework for analysis. *Global Environmental Politics, 9*(4), 14–40.

Brand, U., & Gorg, C. (2013). Regimes in global environmental governance and the internationalization of the state: The case of biodiversity politics. *International Journal of Social Science Studies, 1*(10), 110–122.

Chan, S., Brandi, C., & Bauer, S. (2016). Aligning transnational climate action with international climate governance: The road from Paris. *Review of European Community & International Environmental Law, 25*(2), 238–247.

Chan, S., van Asselt, H., Hale, T., Abbott, K., Beisheim, M., Hoffmann, M., ..., Widerberg, O. (2015, November). Reinvigorating international climate policy: A comprehensive framework for effective nonstate action. *Global Policy*, 1–8.

Ciplet, D., & Roberts, J. (2017). Climate change and the transition to neoliberal environmental governance. *Global Environmental Change, 46*, 148–156.

Cushman, J., Jr., & Banerjee, N. (2018). Urgent climate action required to protect tens of thousands of species worldwide, new research shows.

Inside Climate News. https://insideclimatenews.org/news/17052018/climate-change-animal-extinction-global-warming-habitat-loss-plants-biodiversity-paris-agreement. Accessed May 18, 2018.

Dimitrov, R. S. (2016). The Paris Agreement on climate change: Behind closed doors. *Global Environmental Politics, 16*(3), 1–11.

Eckersley, R. (2004). *The Green State: Rethinking democracy and sovereignty.* Cambridge, MA: The MIT Press.

Eckersley, R. (2005). Greening the nation-state: From exclusive to inclusive sovereignty. In J. Barry & R. Eckersley (Eds.), *The state and the global ecological crisis.* Cambridge: The MIT Press.

Edmondson, B. (2015). Collective responsibilities: A basis for order in the 21st century. In S. Litz (ed.), *Globalization and responsibility* 2015. Champaign, IL: Common Ground Publishing.

Edmondson, B., & Levy, S. (2013). *Climate change and order: The end of prosperity and democracy.* Houndmills: Palgrave Macmillan.

Falkner, R. (2016). The Paris Agreement and the new logic of international climate politics. *International Affairs, 92*(5), 1107–1125.

Galvani, A. P., Bauch, C. T., Anand, M., Singer, B. H., & Levin, S. A. (2016, December 20). Human-environment interactions in population and ecosystem health. *Proceedings of the National Academy of Sciences of the United States of America, 113*(51), 14502–506.

Hempel, L. C. (1996). *Environmental governance the global challenge.* Washington, DC: Island Press.

Hoffman, J., & Hoffman, M. (2008). *Green: Your place in the new energy revolution.* Houndmills: Palgrave Macmillan.

Höhne, N., Kuramochi, T., Warnecke, C., Röser, F., Fekete, H., Hagemann, M., Day, T., …, Gonzales, S. (2017). The Paris Agreement, resolving the inconsistency between global goals and national contributions. *Climate Policy, 17*(1), 16–32. https://doi.org/10.1080/14693062.2016.1218320.

http://unfccc.int/resource/climateaction2020/tep/index.html. Accessed May 25, 2018.

https://climateactiontracker.org/. Accessed May 25, 2018.

https://unfccc.int/process/the-paris-agreement/status-of-ratification. Accessed June 1, 2018.

http://unfccc.int/paris_agreement/items/9444.php. Accessed March 7, 2018.

Karlsson-Vinkhuyzen, S. I., Groff, M., Tamás, P. A., Dahl, A. L., Harder, M., & Hassall, G. (2017). Entry into force and then? The Paris Agreement and state accountability. *Climate Policy.* https://doi.org/10.1080/14693062.2017.1331904.

Kates, R. W., Travis, W. R., & Wilbanks, T. J. (2012, May 8). Transformational adaptation when incremental adaptations to climate change are insufficient. *Proceedings of the National Academy of Sciences of the United States of America, 109*(19), 7156–61.

Keohane, R. O. (2015). The global politics of climate change: Challenge for political science. *Political Science and Politics, 48*(1), 19–26. https://doi.org/10.1017/S1049096514001541.

Keohane, R. O., & Victor, D. G. (2010). *The regime complex for climate change* (Harvard Project on International Climate Agreements Discussion Paper 10–33, 1–30). Cambridge, MA.

Keohane, R. O., & Victor, D. G. (2011). The regime complex for climate change. *Perspectives on Politics, 9*(1), 7–23.

Kinley, R. (2017). Climate change after Paris, from turning point to transformation. *Climate Policy, 17*(1), 9–15. https://doi.org/10.1080/14693062.2016.1191009.

Martin, J., Maris, V., & Simberloff, D. S. (2016, May 31). The need to respect nature and its limits challenges society and conservation science. *Proceedings of the National Academy of Sciences of the United States of America, 113*(22), 6105–12.

Monbiot, G. (2015). Grand promises of Paris climate deal undermined by squalid retrenchments. *The Guardian.* https://www.theguardian.com/environment/georgemonbiot/2015/dec/12/paris-climate-deal-governments-fossil-fuels. Accessed April 27, 2018.

Nelson, D. R. (2009). Conclusions: Transforming the world. In W. N. Adger, I. Lorenzoni, & K. L. O'Brien (Eds.), *Adapting to climate change: Thresholds, values, governance.* Cambridge: Cambridge University Press.

O'Neill, K. (2009). *The environment and international relations.* Port Melbourne, VIC: Cambridge University Press.

Orsini, A., Morin, J., & Young, O. (2013). Regime complexes: A buzz, a boom or a boost for global governance? *Global Governance, 19,* 27–39.

Paterson, M. (2000). *Understanding global environmental politics: Domination, accumulation and resistance.* New York: St. Martin's Press.

Ruggie, J. G. (2014). Global governance and "new governance theory", lessons from business and human rights. *Global Governance, 20,* 5–17.

Sabel, C. F., & Victor, D. G. (2017). Governing global problems under uncertainty, making bottom-up climate policy work. *Climatic Change, 144*(1), 15–27.

Stern, N. (2009). *A blueprint for a safer planet: How to manage climate change and create a new era of progress and prosperity*. London: The Bodley Head.

Stokke, O. S., & Vidas, D. (1996). *Governing the Antarctic: The effectiveness and legitimacy of the Antarctic treaty system*. Cambridge and New York: Cambridge University Press.

Suzuki, D., & Hannington, I. (2017). *Just cool it!: The climate crisis and what we can do, a post-Paris Agreement*. Sydney: Newsouth.

Wettestad, J. (1999). *Designing effective environmental regimes: The key conditions*. Cheltenham: Edward Elgar.

Widerberg, O., & Pattberg, P. (2017). Accountability challenges in the transnational regime complex for climate change. *Review of Policy Research, 34*(10), 68–87.

www.wearestillin.com. Accessed March 7, 2018.

Young, O. (2002). *The institutional dimensions of environmental change: Fit, interplay, and scale*. Cambridge: The MIT Press.

4

Imperatives for Climate Governance for States in the Anthropocene: An Agenda for Transformation

Josephine Mummery and Jane Mummery

1   Governance in the Anthropocene

In the Anthropocene era, governance will increasingly need to differ from that perceived as successful to date. Recognition of the Anthropocene includes acceptance that humans are now the main drivers of planetary-scale climate change, and that the safe operating boundaries of the global environment or Earth system are being exceeded (Pattberg & Widerberg, 2015; Rockström et al., 2009). Without significant changes in human behaviour globally, we face a future of increasing instability in the Earth system, with likely dynamic and damaging consequences, not all of which are known (Steffen et al., 2015). Projections of climate change, for example, describe rates and magnitude of changes

J. Mummery
University of Canberra, Canberra, ACT, Australia

J. Mummery (✉)
Federation University Australia, Ballarat, VIC, Australia
e-mail: j.mummery@federation.edu.au

© The Author(s) 2019
B. Edmondson and S. Levy (eds.), *Transformative Climates and Accountable Governance*, Palgrave Studies in Environmental Transformation, Transition and Accountability, https://doi.org/10.1007/978-3-319-97400-2_4

well outside of natural cycles and variability over the last million years in the climate system in response to anthropogenic greenhouse gas emissions. These will have far-reaching consequences for societies and ecosystems. Recognition, then, of the Anthropocene thus calls for fundamental reconsiderations of the relationships between humans, human societies and the world around us, including the belief systems, institutions and governance approaches that deliver to human needs and aspirations (Hoffman & Jennings, 2015).

Many existing institutions and governance approaches exacerbate the challenges of the Anthropocene, through unsustainable exploitations of ecological resources and ongoing subordination of planetary ecological functioning to short-term extractive and other interests. As Dryzek (2016, p. 941) notes, 'many of the institutions that developed in the Holocene, such as sovereign states and capitalist markets, were complicit in generation of the unstable Earth system that now characterizes the Anthropocene'. The alignment of such institutions with powerful political and economic forces that benefit from material growth underpins institutional path dependencies that preference past practices and maintenance of the status quo. Future choices may thus be constrained by experience increasingly at odds with changes in planetary conditions.

State institutions have contributed both directly and indirectly to the creation of a plethora of global and multilateral institutions, now numbering over 1100, intended to enable solutions to cross-border and global commons environmental challenges such as climate change (Pattberg & Widerberg, 2015). The complexity and scale of Anthropocene challenges make it too early to comprehensively assess the successes of these mechanisms in addressing such issues as climate change. However, there is growing evidence that our actions to, for example, reduce greenhouse gas emissions are not sufficient to avoid critical tipping points, and deeper efforts are needed (Biermann et al., 2012; Pattberg & Widerberg, 2015; Rockström et al., 2009; Sabel & Victor, 2017). It is urgent now to determine and implement reforms of institutions and governance approaches to keep planetary systems clear of irreversible degradation.

The need for transformation of existing systems and processes to effectively mitigate and adapt to climate change is recognised in scholarly and sectoral studies; however, there has been relatively little articulation of how governance approaches themselves need to be restructured to drive wider societal transformations. Institutional and governance studies recognise the magnitude of reforms required and call for further scholarship into a more dynamic institutionalism, trans-disciplinary research and a greater focus on empiricism, co-exploration and co-creation (Dryzek, 2016; Hoffman & Jennings, 2015; Lövbrand et al., 2015; Sabel & Victor, 2017). Normatively, there is also increasing debate over the best understandings of equity and fairness for protecting human—and more-than-human—interests in a finite planet (Gardiner, 2011; Pattberg & Widerberg, 2015; Schlosberg, 2014) (see for instance the discussions developed in Chapter 9 by Martina Grecequet, Jessica J. Hellmann, Jack de Waard and Yudi Li). In similar vein, Dryzek (2016, p. 945) contends, for instance, that our social values need re-thinking so that institutions can transform to better align with dynamic social-ecological contexts, hence stressing the need to 're-think what values such as justice mean in the context of an active and unstable Earth system'.

It is this dynamic complex of present and anticipated socio-ecological conditions, needs and demands for human and more-than-human interests that contextualise this chapter. Our focus here is on detailing what we propose should be the four imperatives driving governance in the Anthropocene:

1. Embedding the primacy of the global environment
2. Transdisciplinary research and knowledge generation
3. Anticipatory and reflexive institutions
4. Learning from and supporting experimentation

These imperatives together provide a framework through which to analyse governance approaches with regard to their capacity to address the challenges of climate change. After elaborating each imperative in turn, we engage them to consider how governance in response to climate change has played out in Australian contexts, notably the renewable energy sector, water policy reform and management of synthetic greenhouse gases. What we aim to foreground through this analysis is that these four imperatives must come to drive governance and policy across all levels and domains (whether directly framed as targeting climate change or not). We argue that polycentric and decentralised approaches and the strong intersectionality of experimentation, anticipation and reflexivity are important for effective governance in the Anthropocene.

2   Imperatives for Governance in the Anthropocene

The Anthropocene demands governance approaches that are capable of responding to interlinked dynamic changes informed with high levels of uncertainty. As detailed below, these imperatives provide the basis for our examinations of climate governance in Australian contexts and final discussions of what they might mean for governance in the Anthropocene.

1. *Embedding the Primacy of the Global Environment*

Embedding the Earth system as a central consideration in governance and decision-making processes must be a core imperative for the Anthropocene. Collectively, global society needs to change course to avoid critical planetary tipping points, and governance mechanisms are key to reducing ecological damage from development processes. This change in course requires a shift away from governance approaches that ignore or at best seek to integrate environmental considerations into human systems, in favour of approaches that consider social and economic systems as operating within the limits set by the biosphere (Hoffman & Jennings, 2015).

Achieving such a shift demands a re-think of the goals and operations of many common institutional and governance approaches that continue to prioritise, for example, protection of sovereignty, economic growth, security and/or human interests over ecological concerns. Global institutions, states, markets and other actors are increasingly considering environmental concerns but typically within a sustainability discourse that does not significantly challenge the high weighting that has conventionally been given to material growth (Biermann, 2012;

Dryzek, 2016; Parr, 2009). Reference to the relative weakness of the environmental pillar in sustainable development as a key global institutional problem, and the capacity of the 2015 UN Sustainable Development Goals to achieve desired progress without a better understanding of how ecosystems contribute to sustainability for all people, has been questioned (Vasseur et al., 2017). While globalisation of research and environmental activism have highlighted the scale of ecological challenges and risks of continued current production and consumption patterns, the globalisation of institutions has also facilitated continued decoupling from local resource limitations and exacerbated problematic path dependencies. Indeed, the interests and functions of the environment are little considered within institutional studies (Hoffman & Jennings, 2015), and emerging environmental problems appear to be addressed through band-aid repairs which do not address underlying human–nature nexus problems (Liu et al., 2015).

Steps towards re-positioning the global environment in institutions and governance approaches can draw on lessons learned to achieve new and clearer conceptualisations of institutional models that are more relevant to the dynamism of the Anthropocene and build reform capacities within institutions. At a global scale, however, examples of successful collective responses to significant environmental challenges are scarce. The 1987 Montreal Protocol for Protection of the Ozone Layer is widely regarded as an effective global environmental response to a potentially catastrophic problem, and Antarctica's governance system is also perceived as effective in achieving its goals (Young, 2011). This does not mean that most environmental agreements are failures; rather, that there are ongoing needs to learn lessons from implementation at all scales and for critical review of processes intended to enable much needed reforms.

Dryzek (2016) identifies 'ecosystemic reflexivity' as the prime directive for institutions in the Anthropocene, involving a dynamic capacity to re-think and change practices on reflection of Earth system responses or early warnings. It calls for governance approaches that are much more aware of the human–nature nexus and its feedbacks, as well as enhanced foresight, experimentation and deliberative practices. The following three governance imperatives can be considered as *enabling* imperatives as they facilitate the realisation of this *core* imperative—embedding the primacy of the environment—through approaches to knowledge generation, institutional reform and experimentation, each of which is tailored to contribute to enhancing ecosystemic reflexivity.

2. *Transdisciplinary Research and Knowledge Generation*

A dynamic and unstable Earth system demands new knowledge as widespread applications of existing knowledge, for example, equilibrium models and tools that assume a static climate, become less valid. Global environmental changes are now anticipated to unfold in more unpredictable ways, as a result of nonlinearities and feedbacks from multiple biogeophysical systems. These changes are also anticipated to reduce the quality and habitability of current environments (Gillings & Hagan-Lawson, 2014). Knowledge generation that aligns with the scope and interlinked nature of planetary-scale change processes is needed to enhance capacities for foresight in the context of these irreversible environmental and climatic changes, thresholds and tipping points.

However, the Anthropocene presents challenges to our knowledge generation systems, with regard to the urgent need for solutions, and the oft-cited gaps between knowledge production and its accessibility and use by decision-makers. The traditional siloed nature of research and conventional assumptions about the primacy of scientific discourses and methods for knowledge generation contribute to these gaps and limit the accessibility of knowledge and ready dialogue between policy-makers and expert knowledge holders. Climate change science is also complex, focuses on timescales beyond the horizon of most decision-making and is embedded with irreducible uncertainty. Uncertainties about climate change impacts, and their long-term nature, mean that conventional 'observe, predict, respond' management approaches are limited in their abilities to inform response strategies (Hulme, 2009). High levels of uncertainty have also been shown to inhibit response action, including actions intended to reduce risks or realise opportunities (Dovers & Hezri, 2010; Weinhofer & Busch, 2013).

Researchers from multiple disciplines must now contribute to more integrated understandings of the Earth system. Described by some as the inter- and transdisciplinary science of coupled Human–Environmental Systems (Seidl et al., 2013), such a science calls for more research into interactions across disciplinary silos, including bio-geophysical, economic, technological, behavioural, cultural and humanist (Pattberg & Widerberg, 2015). Lövbrand et al., (2015) call for the interpretive sciences to engage more actively with Anthropocene research and to continually revisit, challenge and debate the cultural and social assumptions that inform how this science makes sense of, and provides guidance on responses to, changing environmental conditions. Social research that facilitates a diversity of dialogues on the future of the Earth can broaden thinking with regard to possible solutions and help address barriers to knowledge uptake and social mobilisation (Hoffman & Jennings, 2015; Lövbrand et al., 2015). The co-design of research, and participatory processes in its interpretation, can also build wider ownership of knowledge and contribute to society's capacity to anticipate, re-think and act on practices that undermine planetary life-support systems (Hoffman & Jennings, 2015; Moser & Dilling, 2011).

The research community also needs to renew its empirical focus, including critical reflexivity with regard to linking knowledge and action, and its contribution to new decision-support models and tools that reflect dynamic systems and a changing climate (Pattberg & Widerberg, 2015). Research capacities of importance in the Anthropocene span potential thresholds and tipping points in social-ecological systems, early warnings regarding changes in ecological states, and evidence and analysis of outcomes from interventions designed to reduce environmental degradation. Researchers must also examine the human/more-than-human nexus in ways that no longer automatically prioritise reductively understood human interests and explore possibilities of human and more-than-human social and institutional organisation.

3. Anticipatory and Reflexive Institutions

The collective action problems of the Anthropocene require institutions that can facilitate societal transitions to new sets of sustainability-oriented social values. Institutions will need to better reflect the interests of the future and the more-than-human, as well as the (revised) interests of people of the present. Such institutions will benefit from high levels of cooperation, links to transdisciplinary research organisations and mechanisms which anticipate future challenges and allow for reform in response to feedback and learning.

In alignment with the globalisation of environmental activism and the authority and planetary influence exercised by multinational corporations, the global institutional environment has become much more complex in recent decades, with many new international agreements and non-state actors. Heede (2014), for example, found 90 firms (more than half of which were investor-owned) were responsible for more than 60% of all carbon and methane emissions between 1854 and 2010. Novel institutional arrangements have also emerged from non-state actors, such as the Forest Stewardship Council (Chan & Pattberg, 2008), the Carbon Disclosure Project and the C49 Cities Climate Leadership Group (Pattberg & Widerberg, 2015), each bringing new actors and governance mechanisms to the institutional capacity of the Anthropocene. The result, however, can best be described as an architecture that has growing capacity but is fragmented, lacks overall structure and has dispersed authority along with an unproductive bifurcation into state-centric and private institutions (Pattberg & Widerberg, 2015; Zelli & van Asselt, 2013).

Consideration of the effectiveness of current institutions in addressing large-scale environmental issues is also sobering. Global negotiations over climate change, for example, have so far failed to slow the rate of global emissions growth, and a study of more than 340 partnerships agreed upon at the 2002 World Summit on Sustainable Development shows a low level of effectiveness (Pattberg, 2012). The reality that the recent exponential growth in knowledge and capacity has not stemmed risks of further exceedance of planetary boundaries highlights the imperative for institutional reform (Hoffman & Jennings, 2015).

The tasks of institutional reforms to meet the requirements of the Anthropocene are daunting, and there are no simple solutions. Existing institutions need leadership to more comprehensively embed the centrality and constraints of the global environment. Multilateral environmental agreements need to emphasise capacities relevant to ecosystemic reflexivity such as more active links between emerging knowledge, monitoring and evaluation, and core programmes; reflection of planetary boundaries in principles and review plans; and an enhanced focus on integration in inter-agreement forums. Perhaps even more importantly, planetary limits and the context of the Anthropocene need to be reflected in the guiding principles of international or global agreements established to govern economic matters, such as the World Trade Organisation, the Organisation for Economic Cooperation and Development, and agreements on extractive industries such as fisheries in the global oceans. Enhanced coordination between research and resource management or tracking institutions will also be needed, and lessons can be learned from such endeavours as the Intergovernmental Panel on Climate Change, the Millennium Ecosystems Assessment and Future Earth.

A diversified governance system that is rapidly growing, multilevel and fragmented, also calls for greater attention on effectiveness, performance and accountability issues. As well as measurement of performance, a better understanding of often unintended side effects of governance beyond the state is needed (Jamieson & Di Paola, 2016; Pattberg & Widerberg, 2015). Accountability systems have struggled with changing institutional actors, a disjunct between state capacities and global problems, and technological innovations which may be biased towards the interests of global capital (Gill, 1998). Novel systems of accountability will be important, such as those for the Forest Stewardship Council which evolved from market-based accountability responding to the failure of governments to halt deforestation, to cooperation between private and civil society organisations, and now calls for a re-engagement of public actors (Chan & Pattberg, 2008).

4. *Learning from and Supporting Experimentation*

Building on the preceding imperatives, we see this fourth imperative as foregrounding approaches towards transformation and reforms that tackle path dependencies. A reliance on traditional risk frameworks that emphasise reductionist approaches and de-emphasise irreducible uncertainty in many current institutions tends at best to enable incremental improvements (Kunreuther et al., 2013; Travis & Bates, 2014; Wise et al., 2014). The challenges of the Anthropocene, however, require a move from incremental models 'conceived from a systems-in-equilibrium perspective, and implemented in a top-down, predict-and-control manner, to systems-wide transformation that is conceived from a non-equilibrium, non-linear view of systems, and is creative, adaptive, imbued with agency, and implemented through social learning' (Ziervogel, Cowen, & Ziniades, 2016, p. 4).

Such shifts in thinking and governance will be difficult as they face barriers from uncertainties and locked-in institutional, disciplinary and behavioural practices (Kates, Travis, & Wilbanks, 2012). Many societal features, such as economic policies, land-use and resource allocation practices, and assumptions about risk are entrenched and maintained by powerful interests, which can present strong barriers to transformation (O'Brien, 2012). Efforts to drive such transformation, particularly in the scale and breadth required for tackling climate change and avoiding dangerous changes to social-ecological systems, can thus be expected to lead to conflicts, trade-offs and winners and losers at different scales (O'Brien, 2012). And yet, as we have noted, it is just this strong model of transformation that can provide a basis to 'imagine, enact and sustain a transformed world and a way of life that is in balance with the carrying capacity of our earth, and where all life flourishes' (Ziervogel et al., 2016, p. 2).

To be effective, steps towards achievement of such transformations will require support for active experimentation and learning, informed by an ongoing critical questioning and challenging of the underlying assumptions, dominant discourses and current governance of the status quo, and of proposed new approaches. Such foregrounding of creativity and experimentalism over incrementalism is supported by the epistemological pluralism and transdisciplinarity called for by our second imperative. It is fundamental, furthermore, to demonstrate how institutions could transition to new sets of social values, understand how to overcome barriers and new problems, and find new ways to respond to the third imperative's demand for linking lessons learned to institutional systems in wider contexts (Hoffman & Jennings, 2015).

Finally, experimentalism thrives less on consensual models of deliberation (which can tend to support incrementalism and a reduction of exploration) and more on contestatory ones that can confront path dependency and widen the repertoire of proposed courses for action (Avelino, Grin, Pel, & Jhagroe, 2016; Kenis, Bono, & Mathijs, 2016). That is, contestatory models of deliberation—accompanied by demands for critical reflexivity—can enable actors to develop new possibilities from questioning and challenging each other's proposals and arguments, and by holding approaches to consensus open and accountable to additional claims for consideration and inclusion (e.g. Bächtiger, 2011; Mummery, 2017; Pattie, 2008). Finally, here, and pragmatically, opportunities for theoretical and practical experimentation abound and will continue to increase. There are existing needs for proposals for innovative global redistributive programmes, science courts, green courts, protection and support of more-than-human interests and novel possibilities for participation (Jamieson & Di Paola, 2016). Additionally, the processes for experimentation must also be open to needs, interests, actors and institutions that are, as yet, unknown.

3   Role of the State in the Anthropocene

As has been stressed, the scale of the collective action problems we currently face and the proliferation of institutions and actors involved in governance of emerging problems have challenged the roles of states. Authority is now more dispersed as are capacities for taking action, and the operations of global environmental governance actors have increased 'beyond the state' (Jamieson & Di Paola, 2016; Orsini, Morin, & Young, 2013). These changes can suggest a diminishing role for states but it would be premature to underestimate their roles in addressing climate change and other planetary-scale problems. In many ways, the functions of states have increased in alignment with the growing dimensions and scales of environmental challenges, alongside an increasing number of international and regional actors in more complex institutional fields. The role of states thus needs to be pragmatically understood with regard to climate change governance. While they may themselves be critiqued or refashioned, states are currently instrumental in giving effect to global instruments; providing funding and technology through development assistance; significantly framing national cultural and societal values regarding climate change; and incentivising communities to take action. The Paris Accord, for example, widely regarded as a success in climate negotiations and state diplomacy, was conditional on agreement to conditions imposed by powerful states and acceptance that states themselves will determine and deliver their domestic contributions (Dimitrov, 2016; Karlsson-Vinkhuyzen et al., 2018). States thus have important roles in achieving environmental outcomes within and beyond their territorial borders, and in addressing normative concerns, such as legitimacy, accountability and fairness, domestically and in engagement in wider contexts (Pattberg & Widerberg, 2015).

There is also often a separation of public and private governance in the literature, with an assumption that they operate in different dimensions (Pattberg & Widerberg, 2015). While many multinational and private sector agents have escaped accountability by traditional forms of governance, states, NGOs and civil society do have some capacity to encourage behaviour change among corporate actors through collective action and by criticising particular behaviours and rewarding others (Jamieson & Di Paola, 2016). Governments, as well as activist stakeholders and NGOs, have also motivated companies to undertake corporate social responsibility (CSR) measures, including initiatives regarding disclosure of climate change actions, to the extent that CSR is now practised by nearly all publicly traded companies (Littell & Doh, 2015). Collective state agreement to the Montreal Protocol 1987, and subsequent regulation of substances that deplete the Ozone Layer, including trade of those substances, enabled industry leadership in technology change and provided an example of public and private sector action to the benefit of the global environment (Canan, Andersen, Reichman, & Gareau, 2015). (For fuller discussion of the importance of trade in environmentally harmful substances, please see especially Chapter 11 by Jeremy Moss. For broader discussions of the importance of trade in climate change governance, please see also Chapter 5 by Ross Mittiga, Chapter 10 by Timothy Cadman, Tek Maraseni, Hugh Breakey and Hwan-ok Ma, and Chapter 9 by Steve Vanderheiden. For further sustained examination of the complex nexus between states, corporations and territorial jurisdictional boundaries, see Chapter 3 by Matthew Rimmer.)

4 Australia—A Case Study

Australia provides an interesting case study for consideration of governance imperatives in a changing climate. While highly exposed to potential loss and damage from climate extremes and to likely impacts from longer-term variable climate, Australia has remained a relative laggard in terms of climate policy outputs and mitigation ambition since negotiations commenced on the UNFCCC and subsequent global agreements more than 20 years ago. Of note, this position is at odds with community opinion, with, for example, the 2014 Lowy Institute poll showing that some 63% of those surveyed considered that Australia should be demonstrating leadership in reducing greenhouse gas emissions rather than waiting for an international consensus (Lowy Institute, 2014). The magnitude of Australia's pledge at the Paris COP has also been criticised as inadequate by Bernie Fraser, Chairman of the Australian Climate Change Authority, as it failed to recognise Australia having the highest emissions per capita and emission-intensity per GDP in the developed world (Cheung & Davies, 2017).

Australia's national circumstances, including its extreme economic reliance on low-cost fossil fuels for both domestic electricity generation and export earnings, and the embeddedness of these fuels in economic growth and regional development processes are often put forward to explain this reluctance for strong policies and measures to decarbonise energy supply (see, for instance, Crowley, 2017). These circumstances have produced powerful political stakeholders with interests in continued fossil fuel exploitation, which appear to have the capacity to ensure that successive national governments maintain commitments to a carbon-intensive economy, despite consideration of increasingly confident climate change science and growing risks that Australia will be left with stranded assets in the fossil fuel sector as international markets shift to less carbon-intensive options. Such a commitment has continued in the face of federal-state relationships that have been far from cooperative and despite recognition that energy supply is constitutionally a state and territory responsibility and therefore traditionally outside federal policy decision-making (Cheung & Davies, 2017). As we suggest, this paradigm has also underpinned a consistent underinvestment in research on understanding climate change and alternatives to a carbon-intensive economy and society.

In contrast to other states, such as Germany, Norway, Denmark or Britain, Australia's climate change policy is highly politicised and lacking bipartisan support. Australia's political system currently includes what can be described as *veto players*, agents who can prevent departure from the status quo and protect minority interests even when they are not popular. The presence of these veto players in the conservative right of the current federal coalition government, including those holding climate change sceptic views, has constrained progress of domestic policies to align with national risks. Nevertheless, recent experience in Australia in responding to climate extremes that exceed medium-term median climate change projections, and differing political persuasions in state and territory governments in a federal system as well as non-state interests, has generated some innovations in governance at a range of scales that can inform approaches to sustainability transformations and future needs in a transforming climate. We outline these innovations, further considering them in the light of our four imperatives, across the governance and policy contexts of: renewable energy policy and measures; water policy reform; and management of synthetic greenhouse gases. Although these are by no means the only possible contexts for examination, they are productive exemplars in that they illustrate the significance of state-led interventions within an institutional field that spans action at multiple scales, and the need for policy leadership to incentivise responses across multiple sectors and diverse actors and stakeholders.

4.1   Renewable Energy Policy and Measures

Australia has abundant and high-quality renewable energy resources including, for example, the highest average solar radiation of any continent, and studies have found that both solar and wind sources have the capacity for large-scale energy generation free from greenhouse gas emissions (Geoscience Australia & BREE, 2014; Shafiullah, Amanullah, Ali, Jarvis, & Wolfs, 2012). Deployment of renewable energy is driven primarily by the national Renewable Energy Target (RET), established by the Australian Government in 2001 and extended in 2009 to ensure that by 2020 some 45,000 GWh of electricity per year comes from renewable sources (Mey, Diesendorf, & MacGill, 2016). From 2001, Australia's renewable energy capacity increased strongly from 10,650 MW to 19,700 MW, incentivised by the RET and supported by investments of the Clean Energy Finance Corporation (CEFC) and the Australian Renewable Energy Agency (ARENA), and the Solar Cities Program, including for research, innovation and demonstration, and community engagement (Climate Change Authority, 2012; Mey et al., 2016). State and territory governments responded to the national policy settings by supporting renewables and establishing ambitious targets exceeding those of the Australian Government, including the Australian Capital Territory (ACT; 100% by 2050), Victoria (40% by 2025), Queensland and Northern Territory (each 50% by 2030), South Australia (50% by 2025) and New South Wales (net zero emissions by 2050) (Cheung & Davies, 2017). Feed-in tariff schemes introduced by state governments also stimulated deployment of renewable energy, with small-scale residential rooftop photovoltaic (PV) generation totalling about 5 GWh by January 2016 across 1.5 million households (Mey et al., 2016).

The conservative Abbott government elected in 2013, however, reduced federal commitments and support for renewable energy, including placing a ban on ARENA and CEFC funding of key wind projects, cutting research funding and reducing the renewable energy target to 33,000 GWh (Cheung & Davies, 2017; Crowley, 2017). Uncertainty in policy settings has continued under the conservative coalition

Turnbull Government elected in 2015, with several reviews and a current policy framing that prioritises reliability over transformation and is unclear on its commitment to enhanced renewable energy deployment (Jotzo & Mazouz, 2017).

Despite this policy change at the federal level, subnational actions on renewable energy continue, and there is substantial involvement of local government in small-scale renewable energy generation and growth in community renewable energy (CRE) initiatives. Surveys of members of the Coalition for Community Energy in Australia show growth in their members and associates network from 2,721 people in 2011 to 21,000 in 2015 (Mey & Hicks, 2015). Projects supported by CRE groups include two operating community wind farms, Denmark Community Wind (1.6 MW) and Hepburn Wind (4.1 MW), and a community-owned solar farm (Mey & Hicks, 2015). There is, in fact, a diversity of community examples and drivers. The Clean Energy for Eternity community group, for example, established in Tathra, a small community in south-east NSW, went on to fundraise, establish a 30 KW solar farm and provide rooftop solar and wind turbines for 6 surf clubs and 12 Rural Fire Service sheds, solar panels for many local houses, the Tathra Primary School and the local Uniting Church, from a community bulk-buy of panels (Flannery & Gilmour, 2018). Similarly, the BREAZE (Ballarat Renewable Energy and Zero Emissions) community group formed as a result of a groundswell of people meeting after Ballarat's Walk Against Warming rally in 2006. With the mission of protecting and enhancing the natural environment, increasing environmental sustainability within the region by promoting and developing renewable sources of energy, and achieving zero greenhouse gas emissions by 2030 (BREAZE, 2017), this group undertakes a wide range of activities. It too facilitated a community bulk-buy of solar panels, as well as the supply of solar panels and/or solar hot water to not-for-profit community organisations. In 2017, BREAZE was appointed by Sustainability Victoria to establish the Community Power Hub Ballarat. This has seen proposals for further community projects, including for a renewable energy cooperative, a bioenergy plant and the retrofitting of buildings for not only renewable energies but the conversion of waste heat into electricity (Community Power Hub, 2018).

The story of governance, policy reforms and corresponding support for and uptake of renewable energies in Australia is one of only piecemeal success given the lack of consistent national leadership. Nevertheless, it is evident that national policy incentives of 2001–2013 allowed for some experimentation at subnational scales of governance and underpinned both a wide market uptake of more mature renewables technologies and community-led mobilisations of technologies. Without an increase in federal commitments and support for renewable energies, however, it will be increasingly difficult for renewables to continue to innovate and compete in an energy market that favours incumbent technologies and economies of scale (Mey et al., 2016), and this may hamper further growth of community action. Importantly, the imperatives suggest that Australia's lack of consistent investment in innovation and systemic reform in this sector is informed by the pervasive belief across the majority of governance institutions and systems that environmental considerations do not necessitate reform of human social and economic systems and the corresponding paucity of ecosystemic reflexivity. Conversely, due to their locatedness, CRE initiatives are well positioned to recognise the interlinkages between social, economic and environmental systems and can hence achieve community backing for local reform including leverage of investment by shareholding companies into renewable energy technologies and initiatives. To be most effective, such initiatives need to be able to rely on broader policy settings that will facilitate their work and larger-scale innovations to support, for instance, not just technological but infrastructure innovations in this sector. At the same time, broader scale levels of governance need to learn from the local successes of CRE initiatives and partnerships, facilitating both their potential scaling-up and their interconnections. There is, in other words, an identified role for increased support from governance institutions at all levels to remove barriers and facilitate societal and economic actions that would contribute to the decarbonisation of Australia's energy supply.

4.2   Water Policy Reform

The Murray-Darling Basin (MDB) in south-eastern Australia, which supports 40% of the state's irrigated agricultural businesses, faces a significant drying trend due to climate change (CSIRO & BoM, 2015). The climate of the MDB fluctuates between decadal droughts and floods, and the river basin is one of the most variable in the world in terms of streamflow and precipitation (Gergis et al., 2012; Grafton et al., 2014). Extensive engineering works including major storages, weirs, channels and barrages have been developed by the Basin states (notably Queensland, New South Wales, Victoria and South Australia) operating autonomously and highly regulate river flows in the MDB. Most river systems in the MDB have been over-allocated (Marshall, Connell, & Taylor, 2013), the volume of water available to maintain freshwater ecosystems has significantly diminished (Pittock & Finlayson, 2013), and the ecosystem health of 17 of the basin's 23 major rivers has been assessed as poor or very poor (Davies, Harris, Hillman, & Walker, 2010).

In response to concerns about the over-exploitation of water and reports of local ecological crises (Connell, 2007), the Australian Government led a number of institutional and governance reforms with the basin states and the Australian Capital Territory (ACT) aimed at building resilience in the MDB. Initiatives have included establishment of environmental flow targets designed to manage the water to significant ecological assets accompanied by funding to enable water re-allocation from 2003, the adoption of the National Water Initiative (NWI) in 2004 that included commitments to return over-allocated rivers to sustainable levels and strengthened interjurisdictional organisations. The Millennium Drought 2001–2009 challenged progress of these arrangements, with implementation of the environmental components of the NWI assessed as poor (NWC, 2011), and led to more significant reforms under the *Water Act 2007* (Commonwealth of Australia, 2007). The Act gave the Australian Government a greater role in managing the Basin's water, established a new Murray-Darling Basin Authority (MDBA) and sustainable diversion limits for water extraction, provided for water charge and market rules to be developed, and required planning for water management including conservation of ecological assets, with a whole of basin plan developed in 2012 and accredited state plans due July 2019.

While it is too early to fully assess the effectiveness of the governance arrangements and mechanisms established by the *Water Act 2007*, there are a growing number of studies that call into question their effectiveness in the context of a changing climate (Grafton, 2017; Marshall & Alexandra, 2016). Of particular concern, the 2012 Basin Plan did not factor in any specific adjustment for the reductions in water availability due to climate change (Commonwealth of Australia, 2012; Horne, 2014) because climate change impacts were deemed too uncertain (Alexandra, 2017). Detailed climate projections had been delivered to the MDBA by Australia's national science organisation (CSIRO) which highlighted likely median reduction in surface water availability of 11% across the whole basin and 13% in the southern basin, and advice on use of a more extreme scenario in water planning reflecting the conditions of the Millennium drought (Alexandra, 2017; Chiew, Cai, & Smith, 2009). Institutional path dependency has also been identified behind a tendency of decision-makers in the MDB to select and fund engineering interventions to address water management issues, with minimal consideration of risks of disruption to ecological processes (Pittock & Finlayson, 2013). The implications of such path dependence include limits on innovation and possible increased exposure of social-ecological systems to shocks of greater magnitude due to climate change (Marshall & Alexandra, 2016).

With regard, then, to the described governance imperatives, what is evident in this sector so far is that while there has been some limited demonstration of an ecosystemic reflexivity, this still assumes a model of individual (such as individualised drought events) rather than systemic stresses (such as climate change) and is still constrained by (short-term) social and economic interests. That is, because of entrenched path dependencies, governance models have not as yet consistently made environmental—and associated more-than-human—considerations, the basis within which human social and economic decisions must be made. These same path dependencies—and the political weight of existing human use groups—also constrain the anticipatory capacity of MDB institutions and innovative policy reform in this complex sector. Due, however, to ongoing environmental degradation and failures in accountability, government faith in the capacity of market mechanisms to integrate the environment is now being questioned. There are increasing arguments for improved and more transparent science and policy integration, with growing recognition that adequate decision-making must also include tools such as hydro-economic modelling that better reflect the uncertainties and magnitude of risks from climate change. Considering the MDB context, the imperatives also foreground the importance of transdisciplinary imaginative experimentation towards new social values and models for living that not only support more-than-human interests as strongly as human interests but are flexible and resilient enough to meet the requirements of climate change.

4.3   Management of Synthetic Greenhouse Gases

Unlike the preceding policy stories, Australia's management of synthetic greenhouse gases has been progressive and builds directly on effective, proactive and collaborative action taken to protect the stratospheric ozone layer. The scientific uncovering of the link between emissions of CFCs (chlorofluorocarbons), then used ubiquitously in refrigerants and aerosols, for example, and depletion of the ozone layer in 1974, followed by reporting of a hole in the ozone layer above Antarctica in 1985, galvanised coordinated action in Australia and internationally for protection of the ozone layer. The Montreal Protocol, agreed in 1987, is, in addition to its unambiguous achievement of a global environmental outcome, a remarkable instrument with several features that still demonstrate leading practice three decades later (Canan & Reichman, 2002; Rae, 2012). Importantly, the Montreal Protocol was developed when the science remained uncertain, yet negotiators agreed and committed to controlling and phasing out the use of ozone-depleting substances (ODS) based on the precautionary principle, and in-built flexibility allowed for adoption of stricter controls as the science became clearer. The Protocol has also been ratified by all 197 United Nations member states, a world first for any treaty, and mechanisms to support action by developing countries enabled all 142 developing countries to meet the 100% phase-out target for CFCs and halons in 2010 (Rae, 2012).

Synthetic greenhouse gases (SGGs) are often used to replace ODSs, and the Australian Government, in partnership with industry, showed considerable foresight in the 1980s in applying the same regulatory framework to SGGs as to ODSs. The *Ozone Protection and Synthetic Greenhouse Gas Management Act 1989* (Commonwealth of Australia) and associated regulations impose import licences, end-use controls and product stewardship requirements on SGGs with full industry support. With experience in the phase-out of ODS, including in exceeding the requirements of the Montreal Protocol, industry in Australia is now taking action to reduce HFCs (hydrofluorocarbons, both SGGs and ODSs) ahead of the international requirements. This follows further proactive action by the Australian Government in agreement to regulate a domestic phase-down of HFCs in 2016 with commencement in January 2018, ahead of the Montreal Protocol schedule.

Foresight by leading Australian scientists in CSIRO in the 1970s led to the establishment of the most comprehensive atmospheric measurement programme in the world for non-carbon dioxide greenhouse gases, including ODSs, and industry reached out to the science community at that time to build a lasting knowledge partnership that underpins Australia's domestic Montreal Protocol outcomes (Fraser, Pearman, & Derek, 2018). The high quality of the CSIRO atmospheric measurement programme enabled detection of unexplained increases in atmospheric concentrations of a particular HFC, allowing their source as an inadvertent by-product of manufacture to be discovered and cooperative action taken for emissions capture and destruction (Fraser, as cited in van Dijk, 2018a). It also provides the basis for international studies that compare methods to assess emissions and, through declining atmospheric concentrations, the success of the Montreal Protocol.

The management of SGGs in Australia thus aligns well with several of the imperatives we have suggested must drive governance in the Anthropocene. The nature and strength of this regulatory policy approach reflect a determination to achieve the environmental objective, and for it not to be jeopardised by free-rider, deferral or veto behaviours. While the potential availability of alternatives to ODSs and SGGs and the alignment of import/export controls with Australian Government Constitutional responsibilities facilitated development of national policy, considerable leadership was demonstrated by industry in accepting the need for changing technology and in the rapidity of implementation of phase-down measures. Trust between leading individuals in industry and the science community underpinned a lasting industry–science partnership, bridged the often-found gap between science and end-user application, and enabled action by industry and government to commence under conditions of uncertainty (Anderson, as cited in van Dijk, 2018b). Incorporation of SGGs in regulations to address ODSs also reflects a measure of anticipation of future challenges, and the alignment of compliance measures with monitoring and in-built flexibility allows for rapid responses to learning from implementation.

5   Emerging Lessons from Australian Governance

Consideration of these exemplars provides important lessons for achieving effective governance in the Anthropocene. They show, for example, ongoing problems of path dependencies—in values, institutional commitments, research foci, risk management and even desired policy outcomes—with regard to truly setting environmental considerations and ecosystemic reflexivity at the centre of policy decisions. Such path dependencies stifle experimentation and innovation and also reduce the likelihood of achieving necessary changes in global behaviour. These exemplars further show that Anthropocene climate policy must, by its nature, enable both the interconnections of diverse agents, and their capacities to learn from each other, thus foregrounding polycentric and decentralised approaches. Also made visible through these discussions has been the importance of strong trust relationships for such approaches between diverse agents and institutions—even in contexts of contestation. Trust supports alignments of understanding regarding problems and possible solutions, transdisciplinary research and knowledge generation, experimentation informed by critical reflection and anticipation of unforeseen challenges, and transformative reform even in contexts of uncertainty and risk.

The exemplars also highlight that state-level leadership is important pragmatically for enabling the incentivising and scaling-up of commitments and achievements. Much more effort is required to bridge the gap between emerging insights from Anthropocene research, and governance and policy practice at multiple scales and state leadership can facilitate this. The space between researchers and practitioners needs to be opened up within the context of the four imperatives to allow for the emergence of novel possibilities and interconnections. The exemplars illustrate some capacity in Australia; however, more is needed to stimulate the innovation necessary for addressing climate change. Returning finally to the imperatives described, we suggest that together they bring to light points in governance and policy-making where even stated objectives to address the challenges of the Anthropocene and climate change will be led astray by path dependencies and their consequence of insufficient imaginings and experimentation. With the work for governance in the Anthropocene being to foster new imaginings, values, institutions and economies through which both human and more-than-human interests can flourish together in a transforming climate, we believe that our four imperatives support the co-building of a much more sustainable world.

Acknowledgements We would like to thank the editors, Beth Edmondson and Stuart Levy, for the opportunity to contribute to this text and for thus enabling us to further explore how our own differing commitments to theoretical and empirical climate change research can be interwoven.

References

Alexandra, J. (2017). Risks, uncertainty and climate confusion in the Murray-Darling Basin reforms. *Water Economics and Policy, 3*(3), 1650038.

Avelino, F., Grin, J., Pel, B., & Jhagroe, S. (2016). The politics of sustainability transitions. *Journal of Environmental Policy & Planning, 18*(5), 557–567.

Bächtiger, A. (2011). *Contestatory deliberation.* Paper Presented at the Epistemic Democracy Conference, Yale University.

Biermann, F. (2012). Greening the United Nations charter: World politics in the anthropocene. *Environment, 54*(3), 6–17.

Biermann, F., Abbott, K., Andresen, S., Bäckstrand, K., Bernstein, S., Betsill, M. M., …, Gupta, A. (2012). Navigating the anthropocene: Improving earth system governance. *Science, 335*(6074), 1306–1307.

BREAZE. (2017). *BREAZE purpose.* Retrieved from https://www.breaze.org.au/breaze/purpose.

Canan, P., Andersen, S. O., Reichman, N., & Gareau, B. (2015). Introduction to the special issue on ozone layer protection and climate change: The extraordinary experience of building the montreal protocol, lessons learned, and hopes for future climate change efforts. *Journal of Environmental Studies and Sciences, 5*(2), 111–121.

Canan, P., & Reichman, N. (2002). *Ozone connections: Expert networks in global environment governance.* Austin: Greenleaf Publishing.

Chan, S., & Pattberg, P. (2008). Private rule-making and the politics of accountability: Analyzing global forest governance. *Global Environmental Politics, 8*(3), 103–121.

Cheung, G., & Davies, P. J. (2017). In the transformation of energy systems: What is holding Australia back? *Energy Policy, 109,* 96–108.

Chiew, F., Cai, W., & Smith, I. (2009). *Advice on defining climate scenarios for use in Murray-Darling Basin Authority basin plan modelling.* CSIRO Report for the Murray-Darling Basin Authority.

Climate Change Authority. (2012). *Renewable energy target review: Final report.* Melbourne: Climate Change Authority, Australian Government.

Commonwealth of Australia. (2007). *Water Act. Act No. 137 of 2007.* Attorney-General's Department, Canberra.

Commonwealth of Australia. (2012). *The Basin Plan 2012, and The 2012 Basin Plan Explanatory Statement.* Federal Register of Legislation. Retrieved from https://www.legislation.gov.au/Details/F2012L02240/Explanatory%20Statement/.

Community Power Hub. (2018). Ballarat Community Power Hub. Retrieved from http://www.communitypowerhub.com.au/ballarat/.

Connell, D. (2007). *Water politics in the Murray-Darling Basin.* Leichhardt: Federation Press.

Crowley, K. (2017). Up and down with climate politics 2013–2016: The repeal of carbon pricing in Australia. *Wiley Interdisciplinary Reviews: Climate Change, 8*(3).

CSIRO and Bureau of Meteorology (BoM). (2015). *Climate change in Australia information for Australia's natural resource management regions: Technical report.* Melbourne: CSIRO and Bureau of Meteorology.

Davies, P. E., Harris, J. H., Hillman, T. J., & Walker, K. F. (2010). The sustainable rivers audit: Assessing river ecosystem health in the Murray-Darling Basin Australia. *Marine and Freshwater Research, 61*(7), 764–777.

Dimitrov, R. S. (2016). The Paris Agreement on climate change: Behind closed doors. *Global Environmental Politics, 16*(3), 1–11.

Di Paola, M. (2015). Virtues for the anthropocene. *Environmental Values, 24*(2), 183–207.

Dovers, S. R., & Hezri, A. A. (2010). Institutions and policy processes: The means to the ends of adaptation. *Wiley Interdisciplinary Reviews: Climate Change, 1*(2), 212–231.

Dryzek, J. (2016). Institutions for the anthropocene: Governance in a changing earth system. *British Journal of Political Science, 46*(4), 937–956.

Flannery, D., & Gilmour, C. (2018). *Working towards a carbon neutral society: Canberra and the region—Proceedings of the CURF annual forum 2017,* Canberra, ACT: Canberra Urban and Regional Futures, University of Canberra.

Fraser, P. J., Pearman, G. I., & Derek, N. (2018). CSIRO non-carbon dioxide greenhouse gas research. Part 1: 1975–1990. *Historical Records of Australian Science, 28,* 1–13. CSIRO Publishing.

Gardiner, S. M. (2011). Climate justice. In J. Dryzek, R. B. Norgaard, & D. Schlosberg (Eds.), *The Oxford handbook of climate change and society* (pp. 309–322). Oxford: Oxford University Press.

Geoscience Australia, & BREE. (2014). *Australian Energy Resource Assessment.* (2nd ed.). Canberra: Geoscience Australia. Retrieved from https://industry.gov.au/Office-of-the-Chief-Economist/Publications/Documents/GA21797.pdf.

Gergis, J., Gallant, A. J. E., Braganza, K., Karoly, D. J., Allen, K., Cullen, L., ..., McGregor, S. (2012). On the long-term context of the 1997–2009 'Big

Dry' in South-Eastern Australia: Insights from a 206-year multi-proxy rainfall reconstruction. *Climatic Change, 111*(3–4), 923–944.

Gill, S. (1998). European governance and new constitutionalism: Economic and Monetary Union and alternatives to disciplinary neoliberalism in Europe. *New Political Economy, 3*(1), 5–26.

Gillings, M., & Hagan-Lawson, E. (2014). The cost of living in the anthropocene. *Earth Perspectives, 1*, 2. https://doi.org/10.1186/2194-6434-1-2.

Grafton, R. Q. (2017). Water reform and planning in the Murray-Darling Basin, Australia. (Editorial). *Water Economics and Policy, 3*(3), 1702001.

Grafton, R. Q., Pittock, J., Williams, J., Jiang, Q., Possingham, H., & Quiggin, J. (2014). Water planning and hydro-climatic change in the Murray-Darling Basin. *Australia. Ambio, 43*(8), 1082–1092.

Heede, R. (2014). Tracing anthropogenic carbon dioxide and methane emissions to fossil fuel and cement producers 1854–2010. *Climatic Change, 122*, 229–241.

Hoffman, A. J., & Jennings, P. D. (2015). Institutional theory and the natural environment: Research in (and on) the anthropocene. *Organization & Environment, 28*(1), 8–31.

Horne, J. (2014). The 2012 murray-darling basin plan – Issues to watch. *International Journal of Water Resources Development, 30*, 152–163.

Hulme, M. (2009). *Why we disagree about climate change: Understanding controversy, inaction and opportunity.* Cambridge: Cambridge University Press.

Jamieson, D. W., & Di Paola, M. (2016). Political theory for the anthropocene. In D. Held & P. Maffettone (Eds.), *Global political theory* (pp. 254–280). Cambridge: Polity Press.

Jotzo, F., & Mazouz, S. (2017). Will the national energy guarantee hit pause on renewables? *The Conversation.* Retrieved from https://theconversation.com/will-the-national-energy-guarantee-hit-pause-on-renewables-85978.

Karlsson-Vinkhuyzen, S. I., Groff, M., Tamás, P. A., Dahl, A. L., Harder, M., & Hassall, G. (2018). Entry into force and then? The Paris Agreement and state accountability. *Climate Policy, 18*(5), 593–599.

Kates, R. W., Travis, W. R., & Wilbanks, T. J. (2012). Transformational adaptation when incremental adaptations to climate change are insufficient. *Proceedings of the National Academy of Sciences of the United States of America, 109*(19), 7156–7160.

Kenis, A., Bono, F., & Mathijs, E. (2016). Unravelling the (post-)political in transition management: Interrogating pathways towards sustainable change. *Journal of Environmental Policy & Planning, 18*(5), 1–17.

Kunreuther, H., Heal, G., Allen, M., Ednhofer, O., Field, C. B., & Yohe, G. (2013). Risk management and climate change. *Nature Climate Change, 3,* 447–450.

Littell, B., & Doh, J. P. (2015). Corporate social responsibility. In T. C. Lawton & T. S. Rajwani (Eds.), *The Routledge Companion to non-market strategy* (pp. 121–136). Abingdon: Routledge.

Liu, J., Mooney, H., Hull, V., Davis, S. J., Gaskell, J., Hertel, T., ..., Li, S. (2015). Systems integration for global sustainability. *Science, 347*(6225), 1258832.

Lövbrand, E., Beck, S., Chilvers, J., Forsyth, T., Hedrén, J., Hulme, M., ..., Vasileiadou, E. (2015). Who speaks for the future of Earth? How critical social science can extend the conversation on the anthropocene. *Global Environmental Change, 32,* 211–218.

Lowy Institute. (2014). *The Lowy Institute Poll 2014.* Retrieved from https://www.lowyinstitute.org/publications/lowy-institute-poll-2014.

Marshall, G. R., & Alexandra, J. (2016). Institutional path dependence and environmental water recovery in Australia's Murray-Darling Basin. *Water Alternatives, 9*(3), 679.

Marshall, G. R., Connell, D., & Taylor, B. M. (2013). Australia's Murray-Darling Basin: A century of polycentric experiments in cross-border integration of water resources management. *International Journal of Water Governance, 1,* 231–251.

Mey, F., Diesendorf, M., & MacGill, I. (2016). Can local government play a greater role for community renewable energy? A case study from Australia. *Energy Research & Social Science, 21,* 33–43.

Mey, F., & Hicks, J. (2015, June). Community renewable energy in Australia: Exploring its character & emergence in the context of climate change action. In *5th EMES International Research Conference on Social Enterprise* (pp. 1–24). Helsinki.

Moser, S. C., & Dilling, L. (2011). Communicating climate change: Closing the science-action gap. In R. Norgaard, D. Schlosberg, & J. Dryzek (Eds.), *The Oxford handbook of climate change and society* (pp. 161–174). Oxford: Oxford University Press.

Mummery, J. (2017). *Radicalizing democracy for the twenty-first century.* London and New York: Routledge.

National Water Commission (NWC). (2011). *The National Water Initiative— Securing Australia's water future: 2011 assessment.* Canberra: National Water Commission.

O'Brien, K. (2012). Global environmental change II: From adaptation to deliberate transformation. *Progress in Human Geography, 36*(5), 667–676.

Orsini, A., Morin, J. F., & Young, O. (2013). Regime complexes: A buzz, a boom, or a boost for global governance? *Global Governance: A Review of Multilateralism and International Organizations, 19*(1), 27–39.

Parr, A. (2009). *Hijacking sustainability.* Cambridge: Polity Press.

Pattberg, P. H. (Ed.). (2012). *Public-private partnerships for sustainable development: Emergence, influence and legitimacy.* Cheltenham: Edward Elgar.

Pattberg, P., & Widerberg, O. (2015). Theorising global environmental governance: Key findings and future questions. *Millennium, 43*(2), 684–705.

Pattie, J. W. (2008). Arguments-based collective choice. *Journal of Theoretical Politics, 20,* 379–414.

Pittock, J., & Finlayson, C. M. (2013). Climate change adaptation in the Murray-Darling Basin: Reducing resilience of wetlands with engineering. *Australasian Journal of Water Resources, 17*(2), 161–169.

Rae, I. (2012). Saving the ozone layer: Why the montreal protocol worked. *The Conversation.* Retrieved from https://theconversation.com/saving-the-ozone-layer-why-the-montreal-protocol-worked-9249.

Rockström, J., Steffen, W., Noone, K., Persson, Å., Chapin III, F. S., Lambin, E. F., …, Nykvist, B. (2009). A safe operating space for humanity. *Nature, 461*(7263), 472–475.

Sabel, C. F., & Victor, D. G. (2017). Governing global problems under uncertainty: Making bottom-up climate policy work. *Climatic Change, 144,* 15–27.

Schlosberg, D. (2014). Ecological justice for the anthropocene. In M. Wissenburg & D. Schlosberg (Eds.), *Political animals and animal politics* (pp. 75–89). London: Palgrave Macmillan.

Seidl, R., Brand, F. S., Stauffacher, M., Krütli, P., Le, Q. B., Spörri, A., …, Scholz, R. W. (2013). Science with society in the anthropocene. *Ambio, 42*(1), 5–12.

Shafiullah, G. M., Amanullah, M. T. O., Ali, A. S., Jarvis, D., & Wolfs, P. (2012). Prospects of renewable energy – A feasibility study in the Australian context. *Renewable Energy, 39*(1), 183–197.

Steffen, W., Richardson, K., Rockström, J., Cornell, S. E., Fetzer, I., Bennett, E. M., …, Folke, C. (2015). Planetary boundaries: Guiding human development on a changing planet. *Science, 347*(6223), 1259855.

Travis, W. R., & Bates, B. (2014). What is climate risk management? *Climate Risk Management, 1,* 1–4.

van Dijk, S. (2018a). A 30-year retrospective on refrigerant policy. *Climate Control News*. Retrieved from http://www.climatecontrolnews.com.au/ opinion/a-30-year-retrospective-on-refrigerant-policy.

van Dijk, S. (2018b). Ozone protection: Three decades of progress. *Climate Control News*. Retrieved from http://www.climatecontrolnews.com.au/ interviews/ozone-protection-three-decades-of-progress.

Vasseur, L., Horning, D., Thornbush, M., Cohen-Shacham, E., Andrade, A., Barrow, E., ..., Jones, M. (2017). Complex problems and unchallenged solutions: Bringing ecosystem governance to the forefront of the UN sustainable development goals. *Ambio, 46*(7), 731–742.

Weinhofer, G., & Busch, T. (2013). Corporate strategies for managing climate risks. *Business Strategy and the Environment, 22,* 121–144.

Wise, R. M., Fazey, I., Stafford Smith, M., Park, S. E., Eakin, H. C., Archer Van Garderen, E. R. M., & Campbell, B. (2014). Reconceptualising adaptation to climate change as part of pathways of change and response. *Global Environmental Change, 28,* 325–336.

Young, O. R. (2011, December). Effectiveness of international environmental regimes: Existing knowledge, cutting-edge themes, and research strategies. *Proceedings of the National Academy of Sciences, 108*(50), 19853–19860.

Zelli, F., & Van Asselt, H. (2013). Introduction: The institutional fragmentation of global environmental governance: Causes, consequences, and responses. *Global Environmental Politics, 13*(3), 1–13.

Ziervogel, G., Cowen, A., & Ziniades, J. (2016). Moving from adaptive to transformative capacity: Building foundations for inclusive, thriving, and regenerative urban settlements. *Sustainability, 8.* https://doi.org/10.3390/ su8090955.

5

The Empire Strikes Back: Fossil Fuel Companies, Investor-State Dispute Settlement, International Trade, and Accountable Climate Governance

Matthew Rimmer

1   Introduction

There has been much debate about the impact of international trade and investment law upon the environment, biodiversity protection, sustainable development, and climate change. This chapter considers the threat posed by investor-state dispute settlement (ISDS) regimes to a transparent and accountable system of climate governance. In particular, it raises concerns that fossil fuel companies will use investment and trade agreements to lock in fossil investments, at a time at which there should be a decarbonizing of the economy. Part 2 provides a case study of an ISDS matter between Lone Pine Resources Inc. and the Government of Canada in respect of a moratorium on fracking on the St. Lawrence River in Quebec under the *North American Free Trade Agreement* (*NAFTA*) (*Lone Pine Resources Inc.* v. *Government of*

M. Rimmer (✉)
Queensland University of Technology, Brisbane, QLD, Australia
e-mail: matthew.rimmer@qut.edu.au

© The Author(s) 2019
B. Edmondson and S. Levy (eds.), *Transformative Climates and Accountable Governance*, Palgrave Studies in Environmental Transformation, Transition and Accountability, https://doi.org/10.1007/978-3-319-97400-2_5

Canada). Part 3 looks at TransCanada's threat to bring an ISDS action against the Obama administration over its decision to halt the Keystone XL Pipeline. Part 4 considers the treatment of the environment, climate change, trade, and investment during renegotiation of the *NAFTA*. This issue has again come into focus with the renegotiation of the agreement under pressure from President Donald Trump. Part 5 provides an outline of the *Trans-Pacific Partnership* (*TPP*)—focusing in particular upon the text of the Investment Chapter and the ISDS regime. There is a need to ensure that trade and investment agreements do not undermine international climate action—such as that represented by the *Paris Agreement* 2015 (UN Doc FCCC/CP/2015/10/Add.1, 29 January 2016).

ISDS is a mechanism that enables foreign investors to seek compensation from national governments at international arbitration tribunals (UNCTAD, 2014, 2018). There has been a particular focus upon ISDS being used by fossil fuel companies and natural resource entities. This has been a significant issue under the *NAFTA* with its ISDS scheme (NAFTA, 1993). In her prescient 2009 book, *The Expropriation of Environmental Governance*, Kyla Tienhaara foresaw the rise of investor-state dispute resolution of environmental matters. She observed:

> Over the last decade there has been an explosive increase of cases of investment arbitration. This is significant in terms of not only the number of disputes that have arisen and the number of states that have been involved, but also the novel types of dispute that have emerged. Rather than solely involving straightforward incidences of nationalization or breach of contract, modern disputes often revolve around public policy measures and implicate sensitive issues such as access to drinking water, development on sacred Indigenous sites and the protection of biodiversity. (Tienhaara, 2009, p. 1)

In that study, Tienhaara observed that investment agreements, foreign investment contracts, and investment arbitration had significant implications for the protection of the environment. She concluded that 'arbitrators have made it clear that they can, and will, award compensation to investors that claim to have been harmed by environmental regulation' (Tienhaara, 2009, p. 2). She also found that 'some of the cases suggest that the mere threat of arbitration is sufficient to chill environmental policy development' (Tienhaara, 2009, p. 3). Tienhaara was equally concerned by the 'possibility that a government may use the threat of arbitration as an excuse or *cover* for its failure to improve environmental regulation' (Tienhaara, 2009, p. 3). In her view, 'it is evident that arbitrators have *expropriated* certain fundamental aspects of environmental governance from states ... [and] environmental regulation has become riskier, more expensive, and less democratic, especially in developing countries' (Tienhaara, 2009, p. 3).

The Obama administration pushed such issues into sharp relief by advocating sweeping international trade agreements, such as the *TPP* (Geist, 2016; Kelsey, 2010, 2013; Lim, Elms, & Low, 2012; Voon, 2013; Sinclair, Trew, & Mertins-Kirkwood, 2016) and the *Trans-Atlantic Trade and Investment Partnership* (*TTIP*) (Moody, 2013–2014). There was public concern about the impact of such mega-trade deals upon the protection of the environment. Although President Donald Trump has withdrawn the USA from the *TPP* negotiations, the remaining eleven members of the *TPP* have pressed ahead with the agreement in 2018 (White House, 2017).

There has been a spectrum of positions in the debate over ISDS in the context of *NAFTA* and the *TPP*. The Obama administration; a range of multinational companies; and global law firms have advocated for a strong version of ISDS in trade agreements. The Trump administration has been somewhat skeptical about the need for ISDS. Others have called for the reformation of ISDS, with proposals for procedural and substantive reforms, including a push for an Investment Court System in the European Union. By contrast, there are a significant number of legal and economic experts who contend that ISDS is a threat to the rule of law and democracy (Public Citizen, 2016b). According to critics of ISDS, investment arbitration should be abandoned in trade agreements, such as *NAFTA* and the *TPP*.

Professor Gus Van Harten has argued that the multilateral agreement on climate change—the *Paris Agreement* 2015—should be safeguarded against the risk of ISDS claims that target climate change action (Van Harten, 2015). He was concerned about the 'chilling effect' of ISDS actions: 'Faced with risks of uncapped financial liability due to ISDS claims, states may be deterred from implementing measures to fulfill their climate change responsibilities' (Van Harten, 2015). Van Harten warned that 'ISDS poses a risk to climate change measures because 'multinational companies and wealthy foreign nationals have a unique legal right and the financial capacity to bring costly ISDS claims against states without first resorting to domestic courts or tribunals (where they offer justice and are reasonably available) for violations of foreign investor rights' (Van Harten, 2015). He recommended: 'To safeguard against the risk of ISDS claims that frustrate or deter climate change action,' it is suggested that a multilateral climate change agreement should include a broad carve-out from all treaties that allow for ISDS arbitration' (Van Harten, 2015). Such an approach would apply to a range of trade agreements—including *NAFTA* and the *TPP*.

Tienhaara (2017) has argued that there are three options to reform trade and investment agreements better align them with climate change mitigation. The first option is to exclude ISDS provisions. The second option is to prohibit fossil fuel industries from using ISDS. The third option is to carve out all government measures taken in pursuit of international obligations from challenge under ISDS. Tienhaara warns that a lack of reform of ISDS will result in unacceptable delays: 'In a rapidly warming world, we simply cannot afford these delays' (Tienhaara, 2017, p. 22).

2   Lone Pine Resources Inc. v. The Government of Canada

There has been particular disquiet about the use of investor-state clauses to challenge environmental regulations in Canada in light of *NAFTA* (*Huffington Post Canada*, 2013). The dispute between Lone Pine Resources Inc. and the Canadian Government over a fracking moratorium has been an important test case in respect of ISDS.

In 2011, the Quebec National Assembly introduced and passed Bill 18, and placed a moratorium on fracking below the St. Lawrence River in order to allow for a full and timely evaluation of the public health and environmental impacts of such activity. The following year, in 2012, the US energy company Lone Pine Resources Inc. notified the Canadian Government that it would challenge the moratorium on fracking in Quebec's St. Lawrence River under an investment clause Chapter 11 of *NAFTA* (*Lone Pine Resources Inc. v. The Government of Canada*). The full complaint was filed on September 6, 2013 (*Lone Pine Resources Inc. v. The Government of Canada*).

Lone Pine objected to the 'arbitrary, capricious, and illegal revocation of the Enterprise's valuable right to mine for oil and gas under the St. Lawrence River by the Government of Quebec without due process, without compensation, and with no cognizable public purpose' (*Lone Pine Resources Inc. v. The Government of Canada*). The company complained that there had been a lack of consultation by the Quebec Government:

> Between 2006 and 2011, Lone Pine, the Enterprise, and their predecessors expended millions of dollars and considerable time and resources in Quebec to obtain the necessary permits and approvals from the Government of Quebec to mine for oil and gas in the province of Quebec, including beneath the St. Lawrence River. Suddenly, and without any prior consultation or notice, the Government of Quebec introduced Bill 18 into the Quebec National Assembly on May 12, 2011 to revoke all permits pertaining to oil and gas resources beneath the St. Lawrence River without a penny of compensation. (*Lone Pine Resources Inc. v. The Government of Canada*)

The energy company lamented: 'Neither Lone Pine nor the Enterprise were given any meaningful opportunity to be heard, any notice that the Act would be passed, or provided any reason or basis for the outright revocation of the Enterprise's permits relating to oil and gas below the St. Lawrence River' (*Lone Pine Resources Inc. v. The Government of Canada*). The energy company bemoaned the political decision: 'All they were told was that the Act was "a political decision," and that nothing could be done to prevent it from being passed' (*Lone Pine Resources Inc. v. The Government of Canada*).

Lone Pine claimed that 'the moratorium on fracking violated the provision of *NAFTA*'s investment chapter that offers investors a "minimum standard of treatment" and "fair and equitable treatment"' (*Lone Pine Resources Inc.* v. *The Government of Canada*). The company complained that 'Lone Pine and the Enterprise have suffered significant damages as a result of Canada's [alleged] violation of Chapter Eleven of *NAFTA*' (*Lone Pine Resources Inc.* v. The *Government of Canada*).

The company brought this investment action at the same time as Lone Pine sought to restructure itself in bankruptcy (Santo, 2013). Glyn Moody has noted that Lone Pine is really a Canadian firm, being 'a Calgary-based firm' that 'would not have standing as a foreign entity to sue Canada under *NAFTA* but [Lone Pine company president] Granger said it can do so because it is registered in Delaware' (Moody, 2013). On its Web site, Lone Pine is described as an independent oil and gas exploration, development, and production company with operations in Canada within the provinces of Alberta, British Columbia, Quebec, and the Northwest Territories (www.lonepineresources.com). Lone Pine Resources Inc. could well suffer the same fate as Philip Morris—which lost an investor action against Australia over the plain packaging of tobacco products, because its shift of assets from Australia to Hong Kong was considered an abuse of process (Hurst, 2015; Rimmer, 2017).

The Government of Canada has contended that 'the Act is a legitimate measure of public interest that applies indiscriminately to all holders of exploration licences that are located fully or partially in the St. Lawrence River' (Hurst, 2015; Rimmer, 2017). In its view, 'the Act cannot be considered an arbitrary, unfair or inequitable measure' (Hurst, 2015; Rimmer, 2017). Specifically, the government has maintained that 'the measure was enacted by a fundamental democratic institution of Quebec and was preceded by numerous studies that establish that the Act seeks to achieve an important public policy objective, namely, the protection of the St. Lawrence River' (Hurst, 2015; Rimmer, 2017). Moreover, it observed that 'the damages claimed by the claimant are highly exaggerated' (Hurst, 2015; Rimmer, 2017). Furthermore, the Government of Canada noted that 'no representative of the Government of Quebec communicated to the claimant any guarantee, promise or specific assurance that could create legitimate expectations relating to the development of hydrocarbon resources that may be found beneath the St. Lawrence River' (Government of Canada). In any case, the Government of Canada commented that 'passing the Act is a legitimate exercise of the Government of Quebec's police power and, thus, the measure cannot constitute an expropriation' (Government of Canada).

The Governments of Mexico and the USA made submissions on the interpretation of *NAFTA* in August 2017 (Government of Canada).

The dispute is progressing slowly and, as at April 2018, was still unresolved.

Elizabeth May, the leader of the Green Party of Canada, has expressed concerns about investor-state provisions being used to challenge sustainability or environmental protection measures in Canada—such as the action by the US energy company Lone Pine Resources Inc. against Quebec's moratorium on fracking (May, 2013). She observed: 'Such cases represent clear barrier[s] to environmental protection and regulation in Canada' (May, 2013). Her preference was that the *TPP* should not include investor clauses at all. May maintained: 'At minimum, I would insist that any inclusion of investor-state arbitration clauses into the *TPP Free Trade Agreement* include clearly stated exceptions against claims of expropriation for any laws or regulations pertaining to environmental, social, or labour policies that a future government may want to pursue' (May, 2013).

The Canadian champion of the right to water, and leader of the Council of Canadians, Maude Barlow, has long been concerned about the impact of trade and investment agreements upon the environment (Barlow, 2007, 2013; Barlow & Clarke, 2002). In her book, *Blue Future*, Maude Barlow is disturbed by the use of ISDS: 'This "investment arbitration boom" is costing taxpayers billions of dollars and preventing legislation in the public interest' (Barlow, 2013, p. 217). She fears that investment clauses are 'used to gain access to the commons resources of other countries, placing the world's forests, fish, minerals, land, air, and water supplies under direct control of transnational corporations' (Barlow, 2013, pp. 214–215). Barlow maintains that the Lone Pine action is an attack upon Quebec's public management of its water rights (Barlow, 2013).

Martine Châtelain, President of Eau secours!, the Quebec-based coalition for a responsible management of water, argued 'based on the principle of precaution, Quebec government's response to the concerns of its population is appropriate and legitimate' (Sierra Club and Council of Canadians, 2013). The President maintained: 'No companies should be allowed to sue a State when it implements sovereign measures to protect water and the common goods for the sake of our ecosystems and the health of our peoples' (Sierra Club and Council of Canadians, 2013).

Canadian environmental lawyer David Boyd has written upon the need to recognize the right to a clean and healthy environment (Boyd, 2015a, b). In his view, ISDS threatens efforts to improve standards of environmental protection. He has noted that 'advocates of enhanced rights for foreign investors claim that trade deals provide exceptions that allow governments to enact environmental policies' (Boyd, 2016). He warned that such advice was misleading: 'While there is language in trade deals that purports to protect governments' right to regulate, many arbitration panels have ignored or narrowly interpreted these provisions, making them practically useless' (Boyd, 2016). Boyd commented that '[t]ackling the hydra-headed challenge of climate change is already difficult and costly for a fossil-fuel exporting nation like Canada' (Boyd, 2016). He wondered: 'Why ratify trade deals that will make it even harder and more expensive' (Boyd, 2016)?

The departure of Stephen Harper as leader of Canada in 2015, and the new rule of Justin Trudeau, could well lead to a different approach by Canada to fracking, fossil fuels, and climate change. There has been disquiet over Lone Pine Resources Inc. aggressively lobbying the Trudeau Canadian Government over Quebec's fracking ban (Wilt, 2016). It was reported that in April and May 2016, the company lobbied 11 MPs, a policy advisor for the Prime Minister's Office, and the chief of staff for Natural Resources Canada (Wilt, 2016). It remains to be seen what, if any, outcome will result from such political lobbying over the dispute.

The Trudeau Government has caused a great deal of cognitive dissonance, with conflicting views on fossil fuels, energy, and climate change. The Trudeau Government has been a supporter of the Kinder

Morgan Pipeline, against the objections of British Columbia Provincial Government and Indigenous communities (Kassam, 2018; Lukacs, 2018b). But the Trudeau Government has also been an advocate of climate action, carbon pricing, and implementation of the *Paris Agreement* 2015 (Berthiaume, 2018).

Stuart Trew of the Council of Canadians maintained that 'Quebec's moratorium on Fracking is legal and supported strongly by the public' (Sierra Club and Council of Canadians, 2013). He maintained that 'corporate profit should never get in the way of environmental and public health safeguards' (Sierra Club and Council of Canadians, 2013). Stuart Trew insisted: 'It's outrageous to even think that we may have to pay Lone Pine not to drill in the St. Lawrence River' (Sierra Club and Council of Canadians, 2013). Trew contended: 'Trade rules shouldn't be used to appease the whims of dirty oil and gas companies' (Sierra Club and Council of Canadians, 2013).

In 2016, the Sierra Club published a new report by Ben Beachy on trade and the environment, *Climate Roadblocks: Looming Trade Deals Threaten Efforts to Keep Fossil Fuels in the Ground* (Beachy, 2016a). Beachy was concerned that there had been a failure to reform ISDS in light of the dispute between Lone Pine and Canada: 'While the *TPP* and *TTIP* would extend this broad right to thousands of additional foreign investors, neither pact is slated to include meaningful safeguards to prevent fossil fuel firms from following Lone Pine's lead in using it to challenge restrictions on fracking' (Beachy, 2016a, p. 11). Beachy concluded: 'Just as the U.S. begins to transition away from fossil fuels, the *TPP* and *TTIP* would empower an unprecedented number of fossil fuel corporations to follow TransCanada's lead in asking private tribunals to help maintain the crisis-prone status quo' (Beachy, 2016a, p. 23). He observed: 'The fight for climate progress already faces enough obstacles without the additional roadblocks imposed by the *TPP* and *TTIP*' (Beachy, 2016a, p. 23). Beachy maintained: 'Replacing these toxic deals with a new climate-friendly model of trade is an essential component of the growing effort to keep fossil fuels in the ground' (Beachy, 2016a, p. 23).

Ilana Solomon of the Sierra Club observed: 'My right to clean water, clean air, and a healthy planet for my family and community has to come before Lone Pine's right to mine and profit' (Solomon, 2013).

She warned: 'This egregious lawsuit - which Lone Pine Resources must drop - highlights just how vulnerable public interest policies are as a result of trade and investment pacts' (Solomon, 2013). She observed: 'Governments should learn from this and other similar cases and stop writing investment rules that empower corporations to attack environmental laws and policies' (Solomon, 2013). Highlighting the case study of Lone Pine Resources Inc., Solomon has warned against the inclusion of investment clauses in the *TPP*.

In her book, *This Changes Everything*, Naomi Klein (2014) considers the fracking revolution. She expressed concerns about the Lone Pine investor action against the Government of Canada (Klein, 2014). Klein observed: 'As the anti-fossil fuel forces gain strength, extractive companies are beginning to fight back using a familiar tool: the investor protection provisions of free trade agreements' (Klein, 2014, p. 358). She found the claims of Lone Pine to be incredible. Nonetheless, Klein said: 'It's easy to imagine similar challenges coming from any company whose extractive dreams are interrupted by a democratic uprising' (Klein, 2014, p. 359). She was concerned that 'current trade and investment rules provide legal grounds for foreign corporations to fight virtually any attempt by governments to restrict the exploitation of fossil fuels, particularly once a carbon deposit has attracted investment and extraction has begun' (Klein, 2014, p. 359). Klein wondered whether the 'real problem is not that trade deals are allowing fossil fuel companies to challenge governments, it's that governments are not fighting back against these corporate challenges' (Klein, 2014, p. 360).

Naomi Klein and Maude Barlow expressed concern about the impact of investor clauses in the lead up to the Canadian election in 2015 and the international Paris climate talks (Klein & Barlow, 2015). The two warned of the current costs incurred in respect of investor actions under *NAFTA*:

> Canada is currently facing $2.6bn in legal challenges from American corporations under *NAFTA*. Current and past challenges have targeted bans against harmful additives to gasoline and exports of PCBs, and a moratorium on fracking. If a future government wants to reinstate our water

laws or fulfill a commitment to serious fossil fuel reduction that might be agreed to in Paris, *TPP* adds a whole new batch of foreign investors to the current group that already have the right to challenge those laws before a private tribunal. (Klein & Barlow, 2015)

The pair warned that investor clauses can be 'used as a weapon against ambitious climate policy' (Klein & Barlow, 2015). Klein and Barlow said that the *TPP* gives 'foreign corporations the right to directly sue our government for new laws or regulations – whether environmental, health or human rights – that they claim negatively affect their bottom line' (Klein & Barlow, 2015).

3   *TransCanada* v. *The Government of the USA*

There has been a concerted campaign by environmentalists, climate activists, ranchers, and Indigenous communities against the construction of the Keystone XL Pipeline. Environmental groups have argued that the TransCanada investor action highlights the dangers of ISDS in both *NAFTA* and the *TPP* (Page, 2016).

Ilana Solomon, Director of the Sierra Club's Responsible Trade Program, said: 'The rejection of the Keystone XL Pipeline was a huge victory for our climate and for everyone who organized, marched, rallied, and spoke out to stop this polluting project' (Page, 2016). She recommended: 'Congress should reject the toxic *TPP* that would expand these rules to even more polluters' (Page, 2016). In her view: 'We need a new model of trade that puts communities and the environment above corporate profits, not another polluter-friendly trade deal' (Page, 2016).

Jane Kleeb, Director of Bold Nebraska, said: 'A secret court that lets Big Oil sue American taxpayers whenever they do not get their way is a bad deal' (Page, 2016). In a further lament: 'Farmers and ranchers are tired of politicians making secret deals that trade away our property rights' (Page, 2016). Jason Kowalski, 350.org US Policy Director, said:

Climate leadership means keeping fossil fuels in the ground, not signing trade deals that block further climate action. The *TPP* is a major backtrack on the climate progress the Obama administration has made, from rejecting the Keystone XL pipeline on climate grounds to implementing a coal moratorium on public lands. This toxic trade deal would misappropriate power to TransCanada and the rogue fossil fuel industry at the expense of people and planet. (Page, 2016)

Ben Schreiber of Friends of the Earth said: 'The absurdity of the Investor State Dispute Settlement system is exposed by TransCanada's demand of more than $15 billion from the American people for protecting our air and water from their destruction' (Page, 2016).

After much equivocation over many years, President Barack Obama rejected the Keystone XL Pipeline in November 2015 (Goldenberg & Roberts, 2015). He stressed: 'America is now a global leader when it comes to taking serious action on climate change' (Goldenberg & Roberts, 2015). He emphasized: 'Frankly, approving that project would have undercut that global leadership, and that is the biggest risk we face: not acting' (Goldenberg & Roberts, 2015).

In response, environmental groups and climate advocates were ecstatic at the decision: 'President Obama's decision to reject Keystone XL because of its impact on the climate is nothing short of historic — and sets an important precedent that should send shockwaves through the fossil fuel industry' (350.org).

In June 2016, the TransCanada Corporation and TransCanada Pipelines Limited requested arbitration under Chapter 11 of *NAFTA* with the US Government (*TransCanada Corporation and TransCanada Pipelines Limited v. The Government of the United States*, 2016). In its outline of the factual background, the TransCanada Corporation and TransCanada Pipelines Limited made a number of complaints about the rejection of the Keystone XL Pipeline. TransCanada maintained that, at the time Keystone submitted its applications for a Presidential Permit, the US Policy was to expedite approval of energy transmission projects. TransCanada objected to the 'politicization' of the approval process for the Keystone XL Pipeline. TransCanada was concerned about the State Department's seven-year review of Keystone's Presidential

Permit applications for the Keystone XL Pipeline—and the delay and ultimate denial of those applications. TransCanada questioned the State Department's stated reasons for denying the application.

TransCanada makes a number of accusations that the actions of the US Government were unjustified. TransCanada alleges that the US Government unjustifiably delayed processing Keystone's applications for a Presidential Permit for the Keystone XL Pipeline. TransCanada maintains that the government unjustifiably denied Keystone's applications for a Presidential Permit for the Keystone XL Pipeline. TransCanada also maintained that the US Government unjustifiably discriminated against Keystone. TransCanada insisted that it also met the jurisdictional requirements under *NAFTA* and the ICSID Convention.

In addition to the ISDS action, TransCanada also filed a lawsuit against the US Government in the US Federal Court in Houston, Texas (Barron-Lopez, 2016; Beaumont, 2016; TransCanada, 2016). They sought that Obama's denial of approval for the pipeline's construction was 'without legal merit' (Barron-Lopez, 2016; Beaumont, 2016; TransCanada, 2016). TransCanada alleged: 'The Administration's action was contrary to Congress' power under the U.S. Constitution to regulate interstate and international commerce' (Barron-Lopez, 2016; Beaumont, 2016; TransCanada, 2016).

The Obama White House cast doubt on the merits of TransCanada's trade claim and said the US Government had a good track record of winning these kinds of disputes before international trade panels. White House press secretary Josh Earnest commented:

> We never lost a case. I think that is an indication that we have not seen corporations be able to use it effectively to change or alter U.S. law. Our strong record in that venue actually, I think, is a strong argument for precisely why Congress should approve the *TPP*. (Dinan, 2016)

The Obama White House was also confident of prevailing in traditional legal action as well.

The action received significant media attention (Dlouhy, 2016; Dlouhy & Wingfield, 2016). There is also a significant academic commentary on the dispute. Writing in the *American University*

International Law Review, Dillon Fowler argued that TransCanada would not succeed in its action against the Government of the USA (Fowler, 2016). Fowler (2016) observed that legitimate litigation, rather than discrimination, caused the relative delay in the permit approval. Moreover, he noted the usual interpretation of the 'fair and equitable treatment' standard was too narrow (Fowler, 2016). Fowler (2016) stressed that TransCanada's claim would be preempted because three years had elapsed since it first alleged unfair treatment. In his view, investors should be required to exhaust all local remedies before pursuing arbitration.

Senior US Congressional Democrat Sander Levin commented that the TransCanada ISDS claim highlighted problems of legitimacy with ISDS under both *NAFTA* and the *TPP* (Levin, 2016). He commented: 'TransCanada's challenge to the President's decision on the Keystone XL Pipeline through a *NAFTA* investment claim further highlights why we must be certain that the *TPP* trade agreement addresses serious concerns about the ISDS procedures' (Levin, 2016). Levin observed that he made two proposals to the Obama White House about ISDS—but it had not chosen to accept them. He said that he had 'recommended that we negotiate a 'diplomatic screen,' which would have allowed the US and Canadian Governments to agree this was not a case that can be pursued through ISDS' (Levin, 2016). He also recalled that he had 'urged the Administration to negotiate a clarification that "arbitrary" conduct does not amount to a violation of the minimum standard of treatment' (Levin, 2016). Levin warned: 'A full and vigorous public debate is needed to identify problems like this one before the *TPP* agreement is signed' (Levin, 2016).

Lori Wallach, the Director of Public Citizen's Global Trade Watch, warned: 'Canadian corporation TransCanada is skirting our courts and laws to demand that an extrajudicial *NAFTA* investor-state tribunal help it to extract $15 billion from U.S. taxpayers because our government decided an oil pipeline is not good for our nation or the environment' (Public Citizen, 2016a).

Professor Jeffrey Sachs and his colleagues Brooke Guven and Lisa Sachs argued that the TransCanada action highlighted the need to stop the *TPP* (Sachs, Güven, & Sachs, 2016). They noted that the Keystone XL Pipeline 'was a terrible idea, since the pipeline would have supported the development of one of the world's high-carbon energy sources at exactly the time when the world needs to, and has agreed to, decarbonize the world energy system' (Sachs et al., 2016). The academics lamented: 'When government actions infringe on a foreign company's economic interest, however, even in cases where regulatory actions were taken in the public interest, foreign investors like TransCanada can seek damages from the U.S. government for lost profits with some chance of success' (Sachs et al., 2016). They expressed concerns about the chilling effect of ISDS: 'Even the threat of a massive ISDS claim will often be enough to deter governments from introducing regulations to protect its citizens or the environment or even from enforcing existing regulations' (Sachs et al., 2016). The writers concluded: 'Our democracy, and our environment, cannot afford to expand the deeply flawed ISDS system' (Sachs et al., 2016).

Environmental defenders, climate activists, and Indigenous networks highlighted the action by TransCanada as an example of how ISDS posed a threat to climate action—both under *NAFTA* and the *TPP* (Sierra Club, Bold Nebraska, Indigenous Environmental Network, and Friends of the Earth, 2016). The environmental groups warned:

> TransCanada announced its intent to use the (ISDS) system in *NAFTA* to ask a private tribunal of three lawyers to order the U.S. government to pay them more than $15 billion as "compensation" for the pipeline rejection – a decision that spared communities the threat of increased climate disruption and spills of dirty tar sands oil. The (*TPP*) … would extend virtually the same broad rights that TransCanada is claiming to more than 9,000 new foreign-owned firms operating in the U.S., roughly doubling the number of foreign corporations that could follow TransCanada's lead and challenge our environmental protections in unaccountable tribunals. (Sierra Club, Bold Nebraska, Indigenous Environmental Network, and Friends of the Earth, 2016)

The groups warned: 'If the Keystone XL rejection is not immune from investor challenges under trade agreements, it is hard to imagine what environmental policies would be safe, especially if the *TPP* were to pass' (Sierra Club, Bold Nebraska, Indigenous Environmental Network, and Friends of the Earth, 2016). The civil society organizations objected: 'The *TPP* would create a powerful roadblock to environmental and social progress by empowering corporations to demand billions of dollars in compensation for climate and environmental policies' (Sierra Club, Bold Nebraska, Indigenous Environmental Network, and Friends of the Earth, 2016).

Amid concerns that 'these ISDS cases could be detrimental to the sovereignty and rights of Native Nations in the U.S. and First Nations in Canada in the protection of their lands, territories, and peoples,' there have also been concerns about how ISDS impacts upon the rights of Indigenous communities (Sierra Club, Bold Nebraska, Indigenous Environmental Network, and Friends of the Earth, 2016). Tom Goldtooth, Executive Director of Indigenous Environmental Network, claimed: 'The *TPP* trumps the sovereignty rights of our Tribal governments to protect our territorial jurisdictions from potential environmental contamination and exploitation' (Sierra Club, Bold Nebraska, Indigenous Environmental Network, and Friends of the Earth, 2016). He noted: 'The Seven Council Fires of the Great Sioux Nation, the Oceti Sakowin fought long and hard against the TransCanada Keystone XL pipeline, achieving success in its presidential rejection' (Sierra Club, Bold Nebraska, Indigenous Environmental Network, and Friends of the Earth, 2016). Goldtooth emphasized: 'Our network of Native-Indigenous Peoples rejects the investor-state clause of the *TPP* that further sets precedence for private corporations to ignore governmental environmental review processes and to escape financial responsibilities' (Sierra Club, Bold Nebraska, Indigenous Environmental Network, and Friends of the Earth, 2016).

Other Indigenous communities around the Pacific Rim have also been alarmed about the impact of ISDS upon Indigenous rights. Notably, in New Zealand, Maori communities brought an (unsuccessful) action against the *TPP* under the *Treaty of Waitangi*, complaining the impact upon their procedural and substantive rights (Rimmer, 2018).

Ben Beachy of the Sierra Club called for a new model of trade: 'The fight for climate progress already faces enough obstacles without the additional roadblocks imposed by the *TPP* and *TTIP*' (Beachy, 2016b; Beachy, 2017). He insisted: 'Replacing these anachronistic deals with a new climate-friendly model of trade is an essential component of the growing effort to keep fossil fuels in the ground' (Beachy, 2017). Ilana Solomon of the Sierra Club has added: 'If we're really serious about taking on the climate crisis, we have to stop entering into trade agreements that constrain our ability to do so' (Light, 2016).

In January 2017, President Donald Trump signed executive actions to advance the construction of the Keystone XL and Dakota Access oil pipelines (ABC News, 2017). TransCanada suspended its action after the election of President Donald Trump and his decision to support the Keystone XL Pipeline (Lou, 2017). Alberta Premier, Rachel Notley, who has been pro-pipelines, said in a media conference call the *NAFTA* suit's suspension was a 'prudent' move that lets the company 'retain their legal rights' without spending large amounts on fees. The Canadian Government of Justin Trudeau viewed Trump's election as 'positive news' for Keystone XL and the energy industry (Lukacs, 2018a). Environmentalists, such as 350.org's Bill McKibben, remain adamantly opposed to the construction of the Keystone XL Pipeline (Gonzalez & Goodman, 2017). Environmental and landowner groups have brought a lawsuit over the Trump administration's approval of the cross-border permit for the Keystone XL tar sands pipeline (Sierra Club, 2017).

4   NAFTA 2.0

Pursuing an 'America First' Trade Policy, President Donald Trump has demanded a renegotiation of *NAFTA* (Swanson, 2017). He has withdrawn the US Government from both the *TPP* negotiations and the *Paris Agreement* 2015 on climate change. He has also launched a trade action against China under the WTO, alleging breaches of intellectual property. The new US Trade Representative, Robert Lighthizer, has questioned the need for ISDS, saying:

> It's always odd to me when the business people come around and say, 'Oh, we just want our investments protected.'... I mean, don't we all? I would love to have my investments guaranteed. But unfortunately, it doesn't work that way in the market... You either are in the market, or you're not in the market. (Levy, 2017)

The position of the new US President and his US Trade Representative has raised questions about the future of ISDS under *NAFTA*. The Trump administration has opposed the inclusion of climate change in *NAFTA*.

Energy companies have been lobbying aggressively to preserve—or even strengthen—the ISDS regime in *NAFTA*. Among them, Jack Gerard of the Petroleum Institute argued that ISDS provided business certainty:

> In a capital intensive industry like natural gas, we need confidence, we need certainty. Anything that would add to the certainty side of the equation is helpful. ISDS is key to that, in knowing that we can put as much certainty as we can around judicial processes to protect investment. (Dlouhy, 2017)

The industry was concerned that Trump would cancel *NAFTA* altogether or weaken existing ISDS provisions that allow investors to sue countries.

In 2018, a number of Democrat Senators in the US Congress—including Bernie Sanders, Elizabeth Warren, and Kirsten Gillibrand—unveiled a bold 'People First' vision for replacing *NAFTA* in a letter to Donald Trump (Sanders, 2018). The Senators called on the President to eliminate *NAFTA* terms that promote the outsourcing of American jobs: 'The ISDS system and the foreign investor protections it enforces that make it easier and cheaper to outsource jobs must be eliminated' (Sanders, 2018).

The Congressional Progressive Caucus contended that 'a renegotiated *NAFTA* must halt the outsourcing of jobs and pollution by including strong environmental standards with swift and certain enforcement' (Congressional Progressive Caucus A Fair Trade Agenda). The Caucus maintained: 'Trade agreements must contain legally binding obligations for partner nations to adopt, maintain, implement, and strengthen policies to protect our air, water, and climate' (Congressional Progressive Caucus A Fair Trade Agenda).

The Caucus was of the view that ISDS tribunals lacked transparency, accountability, legitimacy, and democratic oversight. The Caucus insisted that there was a need to abolish ISDS under *NAFTA* given their concerns about the impact upon public regulation:

> A renegotiated *NAFTA* must eliminate the special corporate privileges – including the private corporate legal system known as Investor State Dispute Settlement (ISDS) – that help corporations outsource jobs, and empower them to attack environmental and health laws and get large payouts of our tax dollars. Foreign investors use ISDS provisions to sue sovereign nations in private tribunals. ISDS grants corporations special rights to bypass domestic court systems and demand taxpayer compensation for consumer protections, environmental regulations, court decisions, and other government actions that they claim violate their expansive *NAFTA* rights. (Congressional Progressive Caucus A Fair Trade Agenda)

The Caucus was also concerned about the 'chilling effect' of ISDS threats by corporations against governments: 'A foreign investor need not pursue a case to conclusion to achieve the watering down of environmental, health, and other public interest policies, or deter the establishment of new ones' (Congressional Progressive Caucus, A Fair Trade Agenda). They argued that 'the mere threat of an ISDS case against an existing or proposed policy raises the prospect that a government will need to spend millions of dollars in tribunal and legal costs to defend the policy, even if the government might ultimately prevail' (Congressional Progressive Caucus, A Fair Trade Agenda).

Ben Beachy of the Sierra Club supported this effort to reform *NAFTA*:

> Because for too long, trade deals like NAFTA have allowed corporations to evade strong climate policies by simply moving their climate pollution – and jobs – to countries with weaker policies. This climate shell game must

stop. No country should have to fear that climate action will be negated, or that jobs will be lost, due to an antiquated trade deal. It's past time that trade deals support workers and climate progress, not corporate polluters. (Beachy, 2018)

He contended that trade deals such as *NAFTA* needed to be consistent with the *Paris Agreement* 2015.

An alliance of environmental, climate, and human rights groups have called for eight essential changes to *NAFTA* (350.org, 2017). The number one priority is to 'eliminate rules that empower corporations to attack environmental and public health protections in unaccountable tribunals' (350.org, 2017). They maintain: 'Broad corporate rights, including ISDS, must be eliminated from *NAFTA* to safeguard our right to democratically determine our own public interest protections' (350.org, 2017). Second, civil society organizations want strong, enforceable environmental and labor standards in the core of the agreement. Third, they called for safeguards of energy sector regulation. Fourth, they wanted to restrict pollution from cross-border motor carriers. Fifth, they wanted governments to require green government purchasers. Sixth, civil society groups wanted to bolster climate protections by penalizing imported goods made with high climate emissions. Seventh, they wanted governments to prioritize policies that minimize climate pollution. Eighth, the alliance wanted a broad protection for environmental and other public interest policies. In addition to such policy demands, the coalition of civil society groups called for greater transparency, accountability, and public input to the *NAFTA* negotiations.

For its part, the Government of Canada under Justin Trudeau has stressed the importance of the environment and climate change in the *NAFTA* talks. Canada's Prime Minister Trudeau said: 'We are certainly looking for a better level playing field across North America on environmental protections' (Rabson, 2017). Canada's Foreign Affairs Minister, Chrystia Freeland, has vowed to push for progressive revisions of *NAFTA* (Porter, 2017). She has wanted to 'guarantee that the modernized *NAFTA* will not only be an exemplary free-trade deal, it will also be a fair trade deal' (Porter, 2017). She wanted to reform 'the

[ISDS] process, to ensure that governments have an unassailable right to regulate in the public interest' (Freeland, 2017). Canada's Environment Minister, Catherine McKenna, has emphasized that climate change is a priority in talks with the USA in *NAFTA* (The Canadian Press, 2017). She stressed that Trump's 'heated rhetoric' would not stop Canada from seeking to protect its water, air, land, and animals from the impacts of climate change under *NAFTA*.

The Canadian courts have been reluctant thus far to entertain challenges to ISDS rulings (*Canada Attorney General v. Clayton*, 2018). The judiciary, though, has highlighted tensions between domestic environmental standards and rules and the ISDS system of arbitration.

The New Democratic Party's international trade critic Tracey Ramsey warned that 'trying to remove or erase climate change from this agreement will be a red flag to all Canadians' (Meyer, 2017).

The Council of Canadians have called for an end to the ISDS provisions in *NAFTA* (Patterson, 2018). In its submission to the House of Commons, the Council of Canadians commented:

> Chapter 11 is the most emblematic problem of corporate privilege within NAFTA. It is a symbol of everything that is wrong with corporate globalization and Canada must seriously negotiate for its removal. Chapter 11's Investor State Dispute Settlement provisions allow corporations to sue states over changes to legislation or regulations that impact corporate profits, even if the changes are made in the public interest. Under NAFTA and other trade agreements with ISDS provisions, corporations get binding rights to sue governments and obtain financial penalties while citizens have no such rights. (The Council of Canadians, 2017)

The Council of Canadians warned: 'As Canada is seen as the provider of resources within North America, *NAFTA* sets the scene for Canada not being able to live up to its Paris Climate Change Accord commitments' (The Council of Canadians, 2017, pp. 7–8). As Professor Gus van Harten has argued, the multilateral agreement on climate change—the *Paris Agreement* 2015—should be safeguarded against the risk of ISDS claims that target climate change action (Van Harten, 2015).

A 2018 report was published by the Sierra Club, the Council of Canadians, and Greenpeace Mexico on *NAFTA* (Ackerman, Bejar, Laxer, & Beachy, 2018). The report concludes: 'We cannot afford to lock North America's communities into another multi-decade pact that ignores climate change' (Ackerman et al., 2018, p. 44).

5   TPP-11

In addition to *NAFTA*, there has been significant debate over the *TPP*—a proposed trade agreement spanning the Pacific Rim. A center-piece of the *TPP* is Chapter 9 of the agreement, which establishes an ISDS regime. WikiLeaks provided an early glimpse of the Investment Chapter, leaking the draft text in May 2015 (WikiLeaks, Secret Trans-Pacific Partnership Agreement). Julian Assange, WikiLeaks editor, commented:

> The *TPP* has developed in secret an unaccountable supranational court for multinationals to sue states. This system is a challenge to parliamentary and judicial sovereignty. Similar tribunals have already been shown to chill the adoption of sane environmental protection, public health and public transport policies. (Wikileaks Secret Trans-Pacific Partnership Agreement)

WikiLeaks noted that 'the oil giant Chevron [invoked ISDS] against Ecuador in an attempt to evade a multi-billion-dollar compensation ruling for polluting the environment' (WikiLeaks, Secret Trans-Pacific Partnership Agreement). WikiLeaks also warned: 'The threat of future lawsuits chilled environmental and other legislation in Canada after it was sued by pesticide companies in 2008/9' (WikiLeaks, Secret Trans-Pacific Partnership Agreement). WikiLeaks was concerned about the implications of the Investment Chapter of the *TPP*: 'ISDS tribunals are often held in secret, have no appeal mechanism, do not subordinate themselves to human rights laws or the public interest, and have few means by which other affected parties can make representations' (WikiLeaks, Secret Trans-Pacific Partnership Agreement). The revelations about the Investment Chapter received wide attention in quality media (Dorling, 2015).

The White House under President Barack Obama pushed for the inclusion of ISDS settlement in the *TPP*. President Obama obtained a fast track authority from the US Congress to present the final text of the TPP on a take-it-or-leave-it basis. Nonetheless, there was strong resistance from Democrats in the US Congress to the *TPP*. Notably, Democrat Leader Nancy Pelosi was highly critical of President Obama's demands for a fast track authority (Pelosi, 2015). She observed that 'we must prepare our people, our economies and our environment for the future' (Pelosi, 2015). Pelosi was particularly animated about the relationship between trade and climate change: 'The climate crisis presents a challenge to the survival of our planet, but it also presents an opportunity to create a clean energy economy' (Pelosi, 2015). In her view, '[w]e must ensure that trading partners play by the rules and uphold their responsibility to their international obligations' (Pelosi, 2015). Pelosi's position represents a significant rebuff to President Obama's model of trade and the environment.

With the publication of the final text of the *TPP*, the US Trade Representative under President Obama sought to justify the inclusion of the Investment Chapter and ISDS:

> *TPP*'s chapter on Investment strengthens the rule of law in the Asia-Pacific region, deters foreign governments from imposing discriminatory or abusive requirements on American investors, and protects the right to regulate in the public interest. To this end, it ensures that American investors have effective remedies in the event of a breach of their rights, while reforming the (ISDS) system by providing for tools to dismiss frivolous claims and instituting a range of other procedural and substantive safeguards. (US Trade Representative, 2015)

It seems remarkable that ISDS is presented as a solution for 'rule of law' problems, when the regime has been subject to criticism for its failure to respect the 'rule of law' (US Trade Representative, 2015).

The US Trade Representative maintains that 'the chapter includes stronger safeguards to close loopholes and to raise the standards of ISDS above virtually all of the other 3200 plus investment-related agreements in effect around the world' (US Trade Representative, 2015). The US Trade Representative contends that the agreement contains a number of safeguards: 'These include underscoring that countries can regulate in the public interest, including on health, safety, financial stability, and environmental protection; expanding the rules discouraging and dismissing frivolous suits; clarifying that the claimant bears the burden to prove all elements of its claims; allowing governments to issue binding interpretations of the agreement; making proceedings fully open and transparent; and providing for the participation of civil society organizations and others parties not a direct party to the dispute' (US Trade Representative, 2015). The US Trade Representative also argues that the regime provides clarifications of key terms in the agreement.

In the final text of the agreement, there is a wide definition provided for 'investment' (*Trans-Pacific Partnership*, 2015). Article 9.9.3 (d) on performance requirements addresses environmental matters: '(d) Provided that such measures are not applied in an arbitrary or unjustifiable manner, or do not constitute a disguised restriction on international trade or investment, paragraphs 1(b), 1(c), 1(f), 2(a) and 2(b) shall not be construed to prevent a Party from adopting or maintaining measures, including environmental measures: (i) necessary to secure compliance with laws and regulations that are not inconsistent with this Agreement; (ii) necessary to protect human, animal or plant life or health; or (iii) related to the conservation of living or non-living exhaustible natural resources' (*Trans-Pacific Partnership*, 2015).

Article 9.15 of the *TPP* concerns 'Investment and Environmental, Health and other Regulatory Objectives' (*Trans-Pacific Partnership*, 2015). This article provides: 'Nothing in this Chapter shall be construed to prevent a Party from adopting, maintaining or enforcing any measure otherwise consistent with this Chapter that it considers appropriate to ensure that investment activity in its territory is undertaken in a manner sensitive to environmental, health or other regulatory objectives' (*Trans-Pacific Partnership*, 2015).

Section B of Chapter 9 of the *TPP* establishes the mechanism for ISDS. There are a number of measures designed to govern the operation of this mechanism. Article 9.17 looks at consultation and negotiation.

Article 9.18 deals with the submission of a claim to arbitration. Article 9.19 addresses the consent of each party to arbitration. Article 9.20 considers the conditions and limitations on consent of each party. Article 9.21 looks at the selection of arbitrators. Article 9.22 focuses upon the conduct of the arbitration. Article 9.23 examines the transparency of arbitral proceedings. Article 9.25 looks at the governing law. Article 9.26 considers expert reports. Article 9.27 examines consolidation of disputes. Article 9.28 deals with the rules for awards. Article 9.29 addresses the service of documents.

Annex 9.B deals with the nature of expropriation. Clause 3 (b) provides that 'Non-discriminatory regulatory actions by a Party that are designed and applied to protect legitimate public welfare objectives, such as public health, safety and the environment, do not constitute indirect expropriations, except in rare circumstances' (*Trans-Pacific Partnership*, 2015).

Chapter 29 of the *TPP* provides for general exceptions. There is further language here on environmental measures necessary to protect human, animal, or plant life or health.

Senator Elizabeth Warren has been particularly critical of the process and the substance of the negotiations in the TPP. She commented: 'From what I hear, Wall Street, pharmaceuticals, telecom, big polluters and outsourcers are all salivating at the chance to rig the deal in the upcoming trade talks' (Zornick, 2014; See also Warren, 2014). Warren has been a vocal opponent of ISDS (Warren, 2015b). She argued: 'Giving foreign corporations special rights to challenge our laws outside of our legal system would be a bad deal' (Warren, 2015a). In her view, 'If a final *TPP* agreement includes ISDS, the only winners will be multinational corporations' (Warren, 2015a).

In 2015, five US Senators raised concerns about foreign trade tribunals undermining environmental regulations (Franken, 2015). Citing the fracking dispute in Canada, the politicians commented upon the danger of investment clauses diminishing environmental laws: 'As the Administration works to limit fossil fuel emissions both domestically and abroad, ISDS tribunals provide a mechanism to erode environmental safeguards' (Franken, 2015). The US Senators commented: 'We believe that the *TPP* does not provide adequate reforms to ISDS to safeguard environmental protections and could jeopardize the ability of any *TPP* country to enact new policies that would fulfill their international climate commitments' (Franken, 2015). They were concerned that investor clauses would undermine international climate commitments: 'As the Administration looks forward to climate negotiations in Paris and aims to hold countries accountable for climate commitments, it is counterproductive and detractive to endorse a trade provision which gives foreign companies the ability to undercut these international commitments' (Franken, 2015).

The Friends of the Earth has also been concerned about the impact of the trade deal, warning that 'the *TPP* is a potential danger to the planet, subverting environmental priorities, such as climate change measures and regulation of mining, land use, and bio-technology' (Friends of the Earth). The group calls upon Pacific Rim countries to 'reject the proposed *TPP* Investment Chapter that would authorize foreign investors to bypass domestic courts and bring suit before special international tribunals biased in favor of multinationals' (Friends of the Earth, 2013). Erich Pica, president of Friends of the Earth, warned that the *TPP* would 'have a chilling effect on future environmental policies that are desperately needed to address climate change, save ecosystems and protect communities' (Friends of the Earth, 2013).

The climate action network, 350.org, has also objected to the inclusion of an investment clause in the *TPP* (350.org Say No to Corporate Power Grabs). The group warns that 'the TPP will massively boost corporate power at the expense of our climate and environment, human and workers' rights, sovereignty and democracy' (350.org Say No to Corporate Power Grabs). They comment that the 'leaked text reveals that the TPP would empower corporations to directly sue governments in private and non-transparent trade tribunals over laws and policies that corporations allege reduce their profits' (350.org Say No to Corporate Power Grabs). The organization observes that 'Legislation designed to address climate change, curb fossil fuel expansion and reduce air pollution could all be subject to attack by corporations as a result of TPP' (350.org Say No to Corporate Power Grabs). The group is concerned that the fossil fuel industry will rely upon investment clauses to challenge fossil fuel divestment efforts.

Film actor Mark Ruffalo expressed concern that the *TPP* would fuel climate chaos and empower corporate polluters (Ruffalo, 2016). He was particularly concerned about the inclusion of an ISDS regime in the agreement: 'The *TPP* would empower foreign investors to drag the U.S. government to private international arbitration tribunals whenever they claim that our environmental, energy or climate policies violate expansive new *TPP* foreign investor privileges' (Ruffalo, 2016).

Nobel Laureate in Economics, Professor Joseph Stiglitz, has also cited the dispute between Lone Pine Resources Inc. and Canada in his analysis of the impact of the *TPP* upon climate change policy (Stiglitz, 2016). He warned: 'Corporations in carbon-intensive resource extraction and electric utility industries are some of the biggest users of these ISDS mechanisms' (Stiglitz, 2016). Stiglitz noted: 'Currently the American firm Lone Pine is challenging Canada's moratorium on hydrofracking under the St. Lawrence River' (Stiglitz, 2016). He suggested that the *TPP* would expand the scope of the system: 'Unlike *NAFTA*, *TPP* explicitly would extend actionable investor rights to cover government contracts for the "exploration, extraction, refining, transportation, distribution or sale" of government-controlled natural resources like "oil, natural gas, ... and other similar resources"' (Stiglitz, 2016). Stiglitz also highlighted TransCanada's challenge under NAFTA's investor regime to President Barack Obama's decision to reject the Keystone XL Pipeline (Stiglitz, 2016).

In 2012, 100 leading jurists and lawyers led by retired justice, Elizabeth Evatt, wrote an open letter, calling upon the negotiators involved in the *TPP* to reject ISDS (TPP Legal, 2012). Evatt and the jurists were concerned that 'the expansion of this regime threatens to undermine the justice systems in our various countries and fundamentally shift the balance of power between investors, states and other affected parties in a manner that undermines fair resolution of legal disputes' (TPP Legal, 2012). Evatt and company observed that ISDS undermined the rule of law, the judicial process, and democratic decision-making (TPP Legal). The jurists stressed: 'Investment arbitration as currently constituted is not a fair, independent, and balanced method for the resolution of disputes between sovereign nations and private investors' (TPP Legal, 2012). The jurists warned: 'The current regime's expansive definition of covered investments and government actions, the grant of expansive substantive investor rights that extend beyond domestic law, the increasing use of this mechanism to skirt domestic court systems and the structural problems inherent in the arbitral regime are corrosive of the rule of law and fairness' (TPP Legal, 2012).

In a 2014 speech, Chief Justice Robert French of the High Court of Australia raised concerns that the judiciary was not properly consulted in respect of the inclusion of ISDS settlement in trade agreements (French, 2014). His Honour highlighted the impact of ISDS in respect of environmental regulation—including the dispute between Lone Pine Resources Inc. and the Government of Canada (French, 2014, p. 7). Chief Justice French was concerned about the impact of ISDS on the rule of law: 'So far as I am aware the judiciary, as the third branch of government in Australia, has not had any significant collective input into the formulation of ISDS clauses in relation to their possible effects upon the authority and finality of decisions of Australian domestic courts' (French, 2014, p. 15). He was concerned that the issue of ISDS has 'the potential to become larger and it is desirable that it be addressed earlier rather than later' (French, 2014, p. 15). Chief Justice French suggested: 'One approach would be to examine the possibility of including requirements in ISDS provisions in appropriate cases for: prior exhaustion of remedies in domestic courts of the Contracting State; preclusion of any challenge to the decision of a domestic court as constituting a breach of the relevant BIT or FTA provisions; and preclusion of any arbitral decision based upon a rejection of a decision on a question of law of a domestic appellate court binding on lower courts' (French, 2014, p. 15).

In 2016, leading legal and economic experts—including Laurence Tribe, Joseph Stiglitz, and Jeffrey Sachs—wrote a letter to the US Congress, urging politicians to reject the *TPP*, and any trade agreement that included ISDS (Public Citizen, 2016b). The letter warned:

> ISDS grants foreign corporations and investors a special legal privilege: the right to initiate dispute settlement proceedings against a government for actions that allegedly violate loosely defined investor rights to

seek damages from taxpayers for the corporation's lost profits. Essentially, corporations and investors use ISDS to challenge government policies, actions, or decisions that they allege reduce the value of their investments. (Public Citizen, 2016b)

The experts were worried by the proliferation of investment arbitration actions against a wide array of regulatory fields: 'In recent years, corporations have challenged a wide range of environmental, health, and safety regulations, fiscal policies, bans on toxins, denials of permits including for toxic waste dumps, moratoria on extraction of natural resources, measures taken in response to financial crises, court decisions on issues ranging from the scope of intellectual property rights to the resolution of bankruptcy claims, policy decisions on privatizations of prisons and healthcare, and efforts to combat tax evasion, among others' (Public Citizen, 2016b). The experts feared: 'This system undermines the important roles of our domestic and democratic institutions, threatens domestic sovereignty, and weakens the rule of law' (Public Citizen 2016b). The concerted opposition to the *TPP* and ISDS will make it difficult for Obama to conclude the trade deal (Dayen & Grim, 2016; Fulton, 2016).

There has been much academic debate how the trade agreement, generally, and this investment regime in particular will impact upon the environment, biodiversity, and climate change (Rimmer, 2016).

Discussing the *Lone Pine* v. *Canada* case, Meredith Wilensky observed that 'the sheer size of the damage awards being sought demonstrates the substantial financial risk that ISDS can create for countries taking action to protect public health and the environment' (Wilensky, 2015, p. 10698). She warned: 'The *TPP* may obstruct advancement of climate-related policies by creating a risk of liability for measures that negatively affect foreign investments' (Wilensky, 2015, p. 10698). Wilensky argued that there needed to be reforms to the *TPP* to address such concerns about the environment and climate change.

In a comprehensive study of ISDS, Gus van Harten expressed concerns that tribunals did not adopt a position of restraint in the review of legislative and executive decisions (Van Harten, 2013). Tamara Salter has argued that there is a need to require arbitrators to consider international agreements on environmental protection (Salter, 2015, pp. 131–135). She contends: 'As awareness of the *TPP* and *TTIP* grow among citizens and influence politicians and the effects of climate change become more severe, the need to incorporate the efforts of governments and private entities seeking an environmental and climate changes consensus into international economic law and dispute resolution will become more urgent' (Salter, 2015, pp. 131–135).

Valenta Vadi (2015) also wonders whether arbitral tribunals could be reformed to address climate change governance. The scholar contends: 'While arbitral tribunals are not the best forum to adjudicate climate-related disputes, due to their limited mandate and their uneven consideration of environmental concerns in the past, they can contribute to global climate governance' (Vadi, 2015, p. 1351).

Christina Beharry and Melinda Kuritzky have argued that the arbitral tribunal system will become better adapted to dealing with environmental disputes: 'Ultimately, the goal for the arbitral system is to develop the capacity to seriously consider the public policy issues and environmental concerns often at stake while fairly adjudicating the claims of investors harmed by state action' (Beharry & Kuritzky, 2015, p. 430). Unfortunately, this optimism seems misplaced. The investor system seems hostile to public policy regulation in respect of the environment, biodiversity, and climate change.

In a March 2016 study, Jacqueline Wilson expressed concerns about the approach of the *TPP* to the protection of the environment (Wilson, 2016). She observed: 'The number of ISDS cases has expanded exponentially since 2000, with high-profile examples including corporate challenges to anti-smoking legislation in Australia and Uruguay, a ban on hydraulic fracturing in Quebec, a government environmental assessment process in Nova Scotia, and, recently, the U.S. government's decision to block the controversial Keystone XL pipeline' (Wilson, 2016, p. 7). Wilson commented: 'Foreign investors have targeted a broad range of government measures in North America, especially in the areas of environmental protection and natural resource management, that allegedly impaired their investments' (Wilson, 2016, p. 7). She commented: 'Despite this bruising experience, the federal government insists on expanding ISDS in pending international trade agreements, including treaties with the European Union (CETA) and the U.S.-led *TPP*' (Wilson, 2016, p. 7). Wilson concluded that '[c]ritics of ISDS, whose ranks are growing, wonder why the government continues to give private, for-profit arbitrators the power to determine the legitimacy of public policy when we have one of the strongest legal systems in the world, protecting all investors regardless of nationality' (Wilson, 2016, p. 7).

After the withdrawal of the USA from the *TPP* under President Donald Trump, the *TPP* was revived by the remaining eleven countries in 2018 (McDonald, 2018). There has been some further qualification of the ISDS regime. A number of side letters have been signed in respect of ISDS between nation-states. However, the environmental problems previously identified with the *TPP* remain with the latest version of the regime (Rimmer, 2016).

6   Conclusion

In a consideration of challenges to climate regulation, there is an identifiable crisis of accountability and legitimacy around ISDS in the context of *NAFTA* and the *TPP*. The investor arbitration dispute between Lone Pine Resources Inc. and the Government of Canada has become a symbol and a leitmotif of larger concerns about the interaction between trade, investment, the environment, and climate change. Likewise, the action by TransCanada against the Government of the USA under *NAFTA* has raised concerns about fossil fuel companies using ISDS clauses to delay and block climate measures. The *TPP* poses significant threats to the environmental protection of the air, water, and land in the Pacific Rim—particularly through the operation of the ISDS regime. There has been concern that trade agreements, with ISDS clauses, have been used to challenge public regulation, particularly in respect of the environment and climate change. The environmental writer George has warned of the dangers of investment clauses in trade deals: 'Investor-state rules could be used to smash any attempt... to leave fossil fuels in the ground' (Monbiot, 2013). Similarly, Professor Joseph Stiglitz has warned that such agreements would 'significantly inhibit the ability of developing countries' governments to protect their environment from mining and other companies' (Stiglitz, 2013). That is a particularly acute concern for developing countries in the Pacific Rim. Stiglitz has emphasized that there is a need to ensure fairness, equality, and equity in trade and globalization—particularly with respect to environmental outcomes (Stiglitz, 2014). He has emphasized that trade and investment agreements must promote sustainable development in the twenty-first century—rather than runaway climate change (Baker, Jayadev, & Stiglitz, 2017).

Alfred de Zayas—the United Nations Independent Expert on the Promotion of a Democratic and Equitable International Order—has lamented: '[ISDS] is a rather recent and arbitrary construction, a privatized form of dispute settlement that accompanies many international investment agreements' (de Zayas, 2015, p. 10). He has maintained that the ISDS regimes cannot be adequately or sufficiently reformed. Alfred de Zayas has instead recommended the abolition of ISDS regimes. He has instead recommended the creation of an international investment court; state–state dispute settlement; and domestic dispute resolution. The Independent Expert concluded 'that the abolition of ISDS does no injustice to investors, who can still avail themselves of the domestic courts and/or the well-tried mechanism of diplomatic protection' (de Zayas, 2015, p. 21). Moreover, he observed that 'the World Bank offers risk insurance, and this should be factored in as a normal cost of doing business' (de Zayas, 2015, p. 21). Alfred de Zayas commented: 'Notwithstanding the imposition of some necessary limits on the hybrid dogmas of market fundamentalism and the doxology of free trade, investors will continue making handsome profits and, precisely by accepting the principles of transparency, accountability and other reasonable public-oriented regulations, they ensure the continuation of a healthy system of free markets accompanied by sustainable development' (de Zayas, 2015, p. 21). Alfred de Zayas has maintained that there is a need to mainstream human rights in trade and investment law (de Zayas, 2016).

References

350.org and Others. (2017, August 20). Replacing NAFTA: Eight essential changes to an environmentally destructive deal. https://www.sierraclub.org/sites/www.sierraclub.org/files/uploads-wysiwig/NAFTA%20Enviro%20Redlines%20FINAL.pdf.

350.org. Keystone XL—Victory! https://350.org/kxl-victory/.

350.org. Say no to corporate power grabs—Reject the Trans-Pacific Partnership. http://campaigns.350.org/petitions/say-no-to-corporate-power-grabs-reject-the-trans-pacific-partnership.

ABC News. (2017, January 25). Donald Trump Advances North Dakota, Keystone XL Pipeline Projects. *ABC News*. http://www.abc.net.au/news/2017-01-25/trump-to-approve-controversial-pipelines/8210026.

Ackerman, F., Bejar A. A., Laxer, G., & Beachy, B. (2018). *NAFTA 2.0: For people or polluters? A climate Denier's trade deal versus clean energy economy*. https://canadians.org/sites/default/files/publications/report-nafta-people-or-polluters.pdf.

Baker, D., Jayadev, A., & Stiglitz, J. (2017). *Innovation, intellectual property, and development: A better set of approaches for the 21st century*. AccessIBSA. http://ip-unit.org/wp-content/uploads/2017/07/IP-for-21st-Century-EN.pdf.

Barlow, M. (2007). *Blue covenant: The global water crisis and the coming battle for the right to water*. New York: The New Press.

Barlow, M. (2013). *Blue future: Protecting water for people and the planet forever*. Toronto and New York: The New Press.

Barlow, M., & Clarke, T. (2002). *Blue gold: The fight to stop the corporate theft of the world's water*. New York and London: The New Press.

Barron-Lopez, L. (2016, July 1). Pipeline developer is suing Obama to resurrect Keystone XL. *The Huffington Post*. http://huff.to/1Za3npf.

Beachy, B. (2016a). *Climate roadblocks: Looming trade deals threaten efforts to keep fossil fuels in the ground*. Washington, DC: The Sierra Club. http://www.sierraclub.org/sites/www.sierraclub.org/files/uploads-wysiwig/climate-roadblocks.pdf.

Beachy, B. (2016b, May 12). Want to keep fossil fuels in the ground? Stop these trade deals. *The Leap*. https://theleapblog.org/want-to-keep-fossil-fuels-in-the-ground-stop-these-trade-deals/.

Beachy, B. (2017, July 1). The corporation behind Keystone XL just laid bare the TPP's threats to our climate. *The Huffington Post*. http://www.huffingtonpost.com/ben-beachy/the-corporation-behind-ke_b_8931802.html.

Beachy, B. (2018, February 9). Climate-friendly trade? We're not the only ones calling for it. *Compass.* https://www.sierraclub.org/compass/2018/02/climate-friendly-trade-we-re-not-only-ones-calling-for-it.

Beaumont, H. (2016, January 7). TransCanada sues Obama administration over Keystone XL rejection. *Vice News.* https://news.vice.com/article/transcanada-sues-obama-administration-over-keystone-xl-rejection?utm_source=vicenewstwitter.

Beharry, C. & Kuritzky, M. (2015). Going green: Managing the environment through international investment arbitration. *American University Law Review, 30*: 383–430 at 430.

Berthiaume, L. (2018, April 16). Trudeau and Macron promise to double down on climate change fight. *The Canadian Press.* http://www.cbc.ca/news/politics/trudeau-jean-paris-environment-1.4621202.

Boyd, D. (2015a). *The optimistic environmentalist: Progressing towards a greener future.* Toronto: ECW Press.

Boyd, D. (2015b). *Cleaner, greener, healthier: A prescription for stronger Canadian environmental laws and policies.* Vancouver: UBC Press.

Boyd, D. (2016, January 11). Don't let trade deals hamper climate progress. *The Toronto Star.* http://www.thestar.com/opinion/commentary/2016/01/11/dont-let-trade-deals-hamper-climate-progress.html.

The Canadian Press. (2013, November 23). Quebec fracking ban lawsuit: Lone Pine Resources Wants $250 million from Ottawa. *Huffington Post Canada.* http://www.huffingtonpost.ca/2012/11/23/quebec-fracking-ban-lawsuit-lone-pine_n_2176990.html.

The Canadian Press. (2017, August 25). Climate change still a NAFTA priority for Canada. *Maclean's.* http://www.macleans.ca/politics/climate-change-still-a-nafta-priority-for-canada/.

Congressional Progressive Caucus. (2017). A Fair Trade Agenda: Renegotiating NAFTA for Working Families. https://cpc-grijalva.house.gov/a-fair-trade-agenda-renegotiating-nafta-for-working-families/ and https://cpc-grijalva.house.gov/uploads/CPC%20Fair%20Trade%20Agenda.pdf.

The Council of Canadians. (2017, July 17). *Brief on the North American free trade agreement renegotiation to the House of Commons Committee on International Trade,* 5. https://www.ourcommons.ca/Content/Committee/421/CIIT/Brief/BR9162410/br-external/CouncilOfCanadians-e.pdf.

Dayen, D., & Grim, R. (2016, September 8). There's a new front in the battle over the Trans-Pacific Partnership. *The Huffington Post.* http://www.huffingtonpost.com.au/entry/tpp-isds-battle_us_57d030cee4b06a74c9f1da0c.

de Zayas, A. (2015, August 5). Promotion of a democratic and equitable international order, United Nations, 10. http://www.un.org/en/ga/search/view_doc.asp?symbol=A/70/285.

de Zayas A. (2016, September 13). Trade agreements should mainstream human rights, UN News Centre. http://www.un.org/apps/news/story.asp?NewsID=54903#.V9lUAfl97IU.

Dinan, S. (2016, January 7). Keystone pipeline lawsuit and ISDS action threatens Obama's Asian trade deal. *The Washington Times*. http://www.washingtontimes.com/news/2016/jan/7/keystone-pipeline-lawsuit-threatens-obamas-trans-p/#.VpOo_6_WZgE.twitter.

Dlouhy, J. (2016, June 26). TransCanada files $15B NAFTA claim on Keystone XL pipeline. *Bloomberg*. https://www.bloomberg.com/news/articles/2016-06-25/transcanada-files-15b-nafta-claim-on-keystone-xl-rejection.

Dlouhy, J. (2017, September 1). Oil firms that cheered regulatory cuts are quaking on NAFTA. *Bloomberg*. https://www.bloomberg.com/news/articles/2017-09-01/oil-firms-that-cheered-regulatory-rollback-are-quaking-on-nafta.

Dlouhy, J., & Wingfield, B. (2016, January 7). Pacific trade deal opponents say keystone case shows pact's risk. *Bloomberg*. http://bloom.bg/1POY5tk.

Dorling, P. (2015, March 26). WikiLeaks reveals local health and environment rules under threat. *The Sydney Morning Herald*. http://www.smh.com.au/federal-politics/political-news/wikileaks-reveals-local-health-and-environment-rules-under-threat-20150325-1m7y8d.html.

Fowler, D. (2016). Keystonewalled: TransCanada's discrimination claim under NAFTA and the future of investor-state dispute settlement. *American University International Law Review, 31*(1), 103–135.

Franken, A. (2015, May 21). Senators urge U.S. trade representative to protect public health and environmental laws from foreign trade tribunals (press release). https://www.franken.senate.gov/?p=press_release&id=3153.

Freeland, C. (2017, August 14). Address by Foreign Affairs Minister on the Modernization of the North American Free Trade Agreement (NAFTA), Government of Canada. https://www.canada.ca/en/global-affairs/news/2017/08/address_by_foreignaffairsministeronthemodernizationofthenorthame.html?wbdisable=true.

French, R. (2014, July 9), Investor-state dispute settlement—A cut above the courts? Supreme and Federal Courts Judges' Conference. http://www.hcourt.gov.au/assets/publications/speeches/current-justices/frenchcj/frenchcj09jul14.pdf.

The Friends of the Earth. (2013, December 4). Friends of the Earth protest Pacific trade deal. http://www.foe.org/news/blog/2013-12-autumn-agitation-against-pacific-trade-deal#sthash.l7HSTLhz.dpuf.

The Friends of the Earth, The *Trans-Pacific Partnership*. http://www.foe.org/projects/economics-for-the-earth/trade/trans-pacific-partnership#sthash.OddY6Sdg.dpuf.

Fulton, D. (2016, September 7). Prominent scholars decry TPP's "Frontal Attack" on law and democracy. *Common Dreams*.

Geist, M. (2016). The trouble with the TPP', the University of Ottawa. http://www.michaelgeist.ca/2016/01/the-trouble-with-the-tpp-day-1-u-s-blocks-balancing-objectives/.

Goldenberg, S., & Roberts, D. (2015, November 7). Obama rejects Keystone XL pipeline and hails US as leader on climate change. *The Guardian*. https://www.theguardian.com/environment/2015/nov/06/obama-rejects-keystone-xl-pipeline.

Gonzalez, J., & Goodman, A. (2017, March 27). Bill McKibben: Trump may have approved Keystone XL, but people will stop this pipeline again. *Democracy Now!* https://www.democracynow.org/2017/3/27/bill_mckibben_trump_may_have_approved.

Government of Canada, NAFTA—Chapter 11—Investment: Cases filed against the Government of Canada—Lone Pine Resources Inc. The Government of Canada. http://www.international.gc.ca/trade-agreements-accords-commerciaux/topics-domaines/disp-diff/lone.aspx?lang=eng, https://cpc-grijalva.house.gov/a-fair-trade-agenda-renegotiating-nafta-for-working-families/ and https://cpc-grijalva.house.gov/uploads/CPC%20Fair%20Trade%20Agenda.pdf.

Hurst, D. (2015, December 18). Australia wins international legal battle with Philip Morris over plain packaging. *The Guardian*. http://www.theguardian.com/australia-news/2015/dec/18/australia-wins-international-legal-battle-with-philip-morris-over-plain-packaging.

Kassam, A. (2018, April 17). Canada: Trudeau vows to push ahead with pipeline plans in spite of protests. *The Guardian*. https://www.theguardian.com/world/2018/apr/16/canada-trudeau-transcanada-pipeline?CMP=share_btn_tw.

Kelsey, J. (Ed.). (2010). *No ordinary deal: Unmasking the Trans-Pacific partnership free trade agreement*. Wellington: Bridget Williams Books.

Kelsey, J. (2013). *Hidden agendas: What we need to know about the Trans-Pacific Partnership Agreement (TPPA)*. Wellington: Bridget Williams Books Limited.

Klein, N. (2014). *This changes everything: Climate vs. capitalism*. New York: Simon & Schuster.

Klein, N., & Barlow, M. (2015, October 15). Stephen Harper's politics put Canada to shame: Don't be distracted by them. *The Guardian*. http://www. theguardian.com/commentisfree/2015/oct/16/stephen-harper-canada-carbon-climate-change.

Levin S. (2016, January 7). 'Levin statement on transcanada ISDS claim', ways and means committee democrats. https://democrats-waysandmeans.house. gov/media-center/press-releases/levin-statement-transcanada-isds-claim.

Levy, P. (2017, October 23). Critique of NAFTA provision highlights team Trump's misconceptions on investment abroad. *Forbes*. https://www. forbes.com/sites/phillevy/2017/10/23/should-team-trump-encourage-investment-in-mexico/#4fe2dbb170b4.

Light, J. (2016, January 8). Keystone lawsuit illustrates Enviros' big problem With TPP. *Bill Moyers.com*. http://bit.ly/1TKLshS.

Lim, C. L., Elms, D., & Low, P. (Eds.). (2012). *The Trans-Pacific partnership: A quest for a twenty-first century trade agreement*. Cambridge: Cambridge University Press.

Lone Pine Resources. http://www.lonepineresources.com/.

Lou, E. (2017, March 1), TransCanada's $15 billion U.S. Keystone XL NAFTA suit suspended. *Reuters*. https://www.reuters.com/article/us-canada-pipeline-lawsuit/transcanadas-15-billion-u-s-keystone-xl-nafta-suit-suspended-idUSKBN1671W1.

Lukacs, M. (2018a, February 9). Revealed: Trudeau government welcomed oil lobby help for US pipeline push. *The Guardian*. https://www.theguardian.com/environment/true-north/2018/feb/09/trudeau-government-welcomed-oil-lobby-help-for-us-pipeline-push-documents.

Lukacs, M. (2018b, April 17). Who's defending Canada's national interest? First nations facing down a pipeline. *The Guardian*. https://www.theguardian.com/environment/true-north/2018/apr/16/whos-defending-canadas-national-interest-first-nations-facing-down-a-pipeline.

May, E. (2013, January 29). Submission: Environmental assessment of *Trans-Pacific Partnership Free Trade Agreement*, the Green Party of Canada. http://elizabethmaymp.ca/submission-environmental-assessment-tpp.

McDonald, T. (2018, March 8). Asia-Pacific Trade Deal signed by 11 Nations. *BBC News*. http://www.bbc.com/news/business-43326314.

Meyer, C. (2017, August 14). How far will Freeland go to get climate change in NAFTA. *National Observer*. https://www.nationalobserver.com/2017/08/14/news/how-far-will-freeland-go-get-climate-change-nafta.

Monbiot, G. (2013, November 4). A global ban on left-wing politics. *The Guardian*. http://www.monbiot.com/2013/11/04/a-global-ban-on-left-wing-politics/.

Moody, G. (2013, October 4). Canadian-based company sues Canada under NAFTA, saying that fracking ban takes away its expected profits. *TechDirt*. https://www.techdirt.com/articles/20131004/07500724750/canada-hit-with-another-massive-investor-state-dispute-settlement-demand.shtml.

Moody, G. (2013–2014). The TTIP updates. http://blogs.computerworlduk.com/open-enterprise/2013/11/ttip-updates-the-glyn-moody-blogs/index.htm.

Page S. (2016, March 9). Environmental advocates tell Congress: Reject the TPP. *Climate Progress*. http://thinkprogress.org/climate/2016/03/09/3757348/tpp-opposition-letter-to-congress/.

Patterson, B. (2018, February 7). Big oil pushes for NAFTA, council calls for end to energy proportionality and ISDS provisions. *The Council of Canadians*. https://canadians.org/blog/big-oil-pushes-nafta-council-calls-end-energy-proportionality-and-isds-provisions.

Pelosi, N. (2015, June 15). Trade promotion authority on its last legs. *USA Today*. http://www.usatoday.com/story/opinion/2015/06/15/congress-trade-fast-track-tpa-pelosi-column/71270294/.

Porter, C. (2017, August 14). Canada wants a new NAFTA to include gender and indigenous rights. *The New York Times*. https://www.nytimes.com/2017/08/14/world/americas/canada-wants-a-new-nafta-to-include-gender-and-indigenous-rights.html.

Public Citizen. (2016a, January 6). TransCanada demands $15 billion in NAFTA investor-state tribunal for XL Pipeline rejection (press release). https://www.citizen.org/sites/default/files/nafta-isds-keystonexl-statement.pdf.

Public Citizen. (2016b, September 7). 220 + law and economics professors urge Congress to reject the TPP and other prospective deals that include investor-state dispute settlement. http://www.citizen.org/documents/isds-law-economics-professors-letter-Sept-2016.pdf.

Rabson, M. (2017, August 8). Canada's hope to get climate change into NAFTA could prove difficult. *CBC*. http://www.cbc.ca/news/politics/nafta-climate-change-negotiations-1.4238866.

Rimmer, M. (2016). Greenwashing the Trans-Pacific Partnership: Fossil fuels, the environment, and climate change. *Santa Clara Journal of International Law, 14*(2), 488–542.

Rimmer, M. (2017). The chilling effect: Investor-state dispute settlement, graphic health warnings, the plain packaging of tobacco products and the Trans-Pacific Partnership. *Victoria University Law and Justice Journal, 7*(1), 76–93.

Rimmer, M. (2018). The Trans-Pacific Partnership and sustainable development: Access to genetic resources, informed consent, and benefit-sharing. In C. Lawson & K. Adhikari (Eds.), *Biodiversity, genetic resources and intellectual property: Developments in access and benefit sharing* (pp. 151–184). Abingdon, Oxon and New York: Routledge.

Ruffalo, M. (2016, April 4). TPP would fuel climate chaos and empower corporate polluters. *EcoWatch.* https://ecowatch.com/2016/04/04/tpp-fuel-climate-change/.

Sachs, J., Güven, B., & Sachs, L. (2016, July 16). TransCanada lawsuit highlights need to scuttle TPP. *MSNBC.* http://www.msnbc.com/msnbc/op-ed-transcanada-lawsuit-highlights-need-scuttle-tpp.

Salter, T. (2015). Investor-state arbitration and domestic environmental protection. *Washington University Global Studies Law Review, 14,* 131–153.

Sanders, B. (2018, February 2). Progressives call on Trump to fundamentally rewrite NAFTA (press release). https://www.sanders.senate.gov/newsroom/press-releases/progressives-call-on-trump-to-fundamentally-rewrite-nafta and https://www.sanders.senate.gov/download/senate-nafta-letter-to-trump?id=C3E61283-46F3-4004-853A-D497F7539BAA&download=1&inline=file.

Santo, J. (2013, September 25). Lone Pine aims to restructure, raise $100m in bankruptcy, Law 360. http://www.law360.com/articles/475765/lone-pine-aims-to-restructure-raise-100m-in-bankruptcy.

Sierra Club. (2017, November 22). Federal Lawsuit Challenging Keystone XL approval will move forward: Court stops Trump administration from flouting environmental laws (press release). https://www.sierraclub.org/press-releases/2018/01/federal-lawsuit-challenging-keystone-xl-approval-will-move-forward.

Sierra Club and Council of Canadians. (2013, October 2). Lone Pine resources files outrageous NAFTA lawsuit against fracking ban (press release). https://content.sierraclub.org/press-releases/2013/10/lone-pine-resources-files-outrageous-nafta-lawsuit-against-fracking-ban.

Sierra Club, Bold Nebraska, Indigenous Environmental Network, and Friends of the Earth. (2016, March 9). Major environmental, landowner, indigenous groups to congress: Learn from Keystone XL, oppose Trans-Pacific Partnership trade deal. *Sierra Club.* https://content.sierraclub.

org/press-releases/2016/03/major-environmental-landowner-indige-nous-groups-congress-learn-keystone-xl and letter https://www.sierraclub.org/sites/www.sierraclub.org/files/uploads-wysiwig/transcanada-trade-letter.pdf.

Sinclair, S., Trew, S., & Mertins-Kirkwood, H. (Eds.). (2016). *The Trans-Pacific Partnership and Canada: A citizen's guide.* Toronto: Lorimer.

Solomon, I. (2013, March 10). No fracking way: How companies sue Canada to get more resources. *The Huffington Post.* http://www.huffingtonpost.ca/ilana-solomon/lone-pine-sues-canada-over-fracking_b_4032696.html. See also Ilana Solomon of the Sierra Club discussing the Lone Pine dispute and the *TPP.* http://vimeo.com/79027080.

Stiglitz, J. (2013, November 9). Developing countries are right to resist restrictive trade agreements. *The Guardian.* http://www.theguardian.com/business/2013/nov/08/trade-agreements-developing-countries-joseph-stiglitz.

Stiglitz, J. (2014, March 15). On the wrong side of globalization. *The New York Times.* http://opinionator.blogs.nytimes.com/2014/03/15/on-the-wrong-side-of-globalization/.

Stiglitz, J. (2016, March 28). The TPP's hidden climate costs. *Roosevelt Institute.* http://rooseveltinstitute.org/tpps-hidden-climate-costs/.

Swanson, A. (2017, July 17). Trump administration unveils goals in renegotiating NAFTA. *The Washington Post.* https://www.washingtonpost.com/news/wonk/wp/2017/07/17/trump-administration-outlines-goals-for-nafta-re-write/?noredirect=on&utm_term=.36fac89d6208.

Tienhaara, K. (2009). *The expropriation of environmental governance: Protecting foreign investors at the expense of public policy.* Cambridge: University of Cambridge Press.

Tienhaara, K. (2017). Regulatory chill in a warming world: The threat to climate policy posed by investor-state dispute settlement. *Transnational Environmental Law,* 1–22.

TPP Legal. (2012, May 8). An open letter from lawyers to the negotiators of the Trans-Pacific Partnership urging the rejection of investor-state dispute settlement. http://tpplegal.wordpress.com/open-letter/.

TransCanada. (2016, January 6). TransCanada commences legal actions following Keystone XL Denial (press release). https://www.transcanada.com/en/announcements/2016-01-06transcanada-commences-legal-actions-following-keystone-xl-denial/.

United Nations Conference on Trade and Development (UNCTAD). (2014, April). Recent Developments in Investor-State Dispute Settlement: Updated for the Multilateral Dialogue on Investment. http://unctad.org/en/ PublicationsLibrary/webdiaepcb2014d3_en.pdfS.

United Nations Conference on Trade and Development (UNCTAD). (2018). ISDS navigator. http://investmentpolicyhub.unctad.org/ISDS.

The United States Trade Representative. (2015, November 5). Chapter summary of Chapter 9 on investment in the *Trans-Pacific Partnership*. https:// medium.com/the-trans-pacific-partnership/investment-c76dbd892f3a (Post removed after new Trump Administration took charge).

Vadi, V. (2015). Beyond known worlds: Climate change governance by arbitral tribunals. *Vanderbilt Journal of Transnational Law, 48,* 1285–1351.

Van Harten, G. (2013). *Sovereign choices and sovereign constraints: Judicial restraint in investment treaty arbitration* (p. 92). Oxford: Oxford University Press.

Van Harten, G. (2015, September 20). *An ISDS carve-out to support action on climate change* (Osgoode Hall Law School Legal Studies Research Paper). https://canadians.org/sites/default/files/publications/VanHarten-EN-Mar2016.pdf.

Voon, T. (Ed.). (2013). *Trade liberalisation and international co-operation: A legal analysis of the Trans-Pacific Partnership agreement.* Cheltenham and Northampton, MA: Edward Elgar.

Warren, E. (2014). *A fighting chance.* New York: Metropolitan Books.

Warren, E. (2015a, February 25). The Trans-Pacific Partnership clause everyone should oppose. *The Washington Post.* https://www.washingtonpost.com/ opinions/kill-the-dispute-settlement-language-in-the-trans-pacific-partnership/2015/02/25/ec7705a2-bd1e-11e4-b274-e5209a3bc9a9_story.html.

Warren, E. (2015b, February 26). The Trans-Pacific Partnership and investor-state dispute settlement. https://www.youtube.com/watch?v=xzfxv2XQoPg.

White House. (2017, January 23). Presidential memorandum regarding withdrawal of the United States from the Trans-Pacific Partnership negotiations and agreement. https://www.whitehouse.gov/presidential-actions/presidential-memorandum-regarding-withdrawal-united-states-trans-pacific-partnership-negotiations-agreement/.

WikiLeaks, Secret Trans-Pacific Partnership Agreement (TPP)—Investment Chapter. https://wikileaks.org/tpp-investment/press.html.

Wilensky, M. (2015). Reconciling international investment law and climate change policy: Potential liability for climate measures under the

Trans-Pacific Partnership. *Environmental Law Reporter News and Analysis*, *45*, 10683–98 at 10684.

Wilson, J. (2016, March). Bait-and-switch—The Trans-Pacific Partnership's promised environmental protections do not deliver. Canadian Centre for Policy Alternatives. https://www.policyalternatives.ca/publications/reports/bait-and-switch.

Wilt, J. (2016, May 25). Lone Pine, company suing Canada over Quebec's fracking ban, aggressively lobbying in Ottawa. *Desmog Canada*. http://www.desmog.ca/2016/05/25/lone-pine-company-suing-canada-quebec-fracking-ban-aggressively-lobbying-ottawa.

Zornick, G. (2014, May 15). Elizabeth Warren reveals inside details on trade talks. *The Nation*. http://www.thenation.com/blog/179885/elizabeth-warren-reveals-inside-details-trade-talks.

Cases

Canada (Attorney General) v. Clayton. (2018). FC 436 (CanLII). http://canlii.ca/t/hrts9 (Mactavish J.).

Lone Pine Resources Inc. v. The Government of Canada. UNCITRAL. http://www.italaw.com/cases/1606.

Submission of Mexico in *Lone Pine Resources Inc. v. Government of Canada*. (2017, August 16). https://www.italaw.com/sites/default/files/case-documents/italaw9272.pdf.

Submission of the United States of America in *Lone Pine Resources Inc. v. Government of Canada*. (2017, August 16). https://www.italaw.com/sites/default/files/case-documents/italaw9271.pdf.

TransCanada Corporation and TransCanada PipeLines Limited v. The Government of the United States of America, Request for Arbitration. (2016, June 24). https://www.state.gov/documents/organization/259329.pdf.

International Treaties

North American Free Trade Agreement 32 ILM 289, 605. (1993). Entry into force 1 January 1994.

Paris Agreement to the United Nations Framework Convention on Climate Change, opened for signature 12 December 2015 (entered into force 4

November 2016) (in UNFCCC, Report of the Conference of the Parties on its Twenty-First Session, Addendum, UN Doc FCCC/CP/2015/10/Add.1, 29 January 2016).

The Trans-Atlantic Trade and Investment Partnership, http://www.ustr.gov/ttip.

The Trans-Pacific Partnership. (2015). https://ustr.gov/trade-agreements/free-trade-agreements/trans-pacific-partnership/tpp-full-text.

The Trans-Pacific Partnership. http://www.ustr.gov/tpp and https://www.dfat.gov.au/fta/tpp/.

6

The Public's Perception of International Climate Leadership: Insights from the European Union

Jale Tosun, Mile Mišić and Nicole M. Schmidt

1 Introduction

Since the beginning of the European integration project with the sig-nature of the Treaty of Rome in 1957, the European Union (EU) has developed in 'scope and significance without parallel among interna-tional organizations' (Moravcsik, 2005, p. 349). The integration attain-ments comprise both the creation of new institutions and the number of policy areas for which the member states agreed to delegate legal competences to the EU. Furthermore, the EU began to shape not only the policy arrangements of its member states, but also those of

J. Tosun (✉) · M. Mišić · N. M. Schmidt
Institute of Political Science, Heidelberg University, Heidelberg, Germany
e-mail: jale.tosun@ipw.uni-heidelberg.de

M. Mišić
e-mail: mile.misic@ipw.uni-heidelberg.de

N. M. Schmidt
e-mail: nicole.schmidt@ipw.uni-heidelberg.de

© The Author(s) 2019   **119**
B. Edmondson and S. Levy (eds.), *Transformative Climates and Accountable Governance*, Palgrave Studies in Environmental Transformation, Transition and Accountability, https://doi.org/10.1007/978-3-319-97400-2_6

non-European countries. The corresponding literature on the external governance of the EU has shown that the prospect of membership and economic and financial resources gives the EU considerable power in affecting environmental policy arrangements in non-EU countries (e.g. Schulze & Tosun, 2013, 2016; Tosun & Schulze, 2015). Many scholars support the view that the EU now plays a central role in establishing and diffusing global environmental norms (e.g. Jordan, Wurzel, Zito, & Brückner, 2003; Kelemen, 2010; Kelemen & Vogel, 2010; Vogel, 2003).

From the early 1990s onwards, the EU has proclaimed a desire to lead the international community in combatting climate change (Huitema et al., 2011; Oberthür & Roche Kelly, 2008; Selin & Van Deveer, 2015). This occurred about the time when climate change became increasingly recognized and was put on the global policy agenda by the United Nations. Most importantly, in 1992, the United Nations Framework Convention on Climate Change (UNFCCC) was formed, which also resulted in the creation of the UNFCCC secretariat in the same year.

The EU, as an independent signatory to the Convention, has been successful in performing the role of an international leader in climate politics. For example, the EU succeeded in bringing about Russia's ratification of the Kyoto Protocol in 1997. The European representatives maintained that they would only support Russia's membership in the World Trade Organization if the government ratified the Kyoto Protocol. This is only one example that demonstrates that the EU—when perceived as a collective actor—is committed to promoting climate policy both internally towards its member states and externally towards international organizations or non-European states. On the one hand, this may be surprising given the complexity of the EU's institutional structure, which is widely referred to as a multilevel system. On the other hand, the EU's multilevel system and the possibility to practise venue shopping can be regarded as one of the reasons for the EU's policymaking activities.

While the EU's multilevel system may be conducive to policymaking, it is conceivable that it blurs competences and therefore makes it difficult to attribute responsibility. The European Commission represents the EU in international negotiations, but the individual EU member states also participate in these negotiations (Oberthür & Gehring, 2006). This is one example of why citizens may find it difficult to attribute responsibility for political action in the EU's multilevel system. Another potential explanation for citizens experiencing difficulty in attributing responsibility is that because individual member states are responsible for implementing EU law (Egeberg & Trondal, 2016), their citizens may not be aware that the policy measures being implemented originate from the European level. In federal polities across the EU, responsibilities for policy implementation are delegated to regional or local levels, which again seems likely to obfuscate the European nature of the policy measures in question.

In terms of accountability, such a blurring of policymaking competences can be problematic from the viewpoint of democracy and needs to be addressed by measures that enhance transparency. Especially when considering that the EU is increasingly active in international climate politics, the question arises whether the EU actually holds a mandate from the public to do so. Otherwise, the EU may promote the 'right' policies with a view to combat climate change, but lack the legitimacy for this action. Therefore, in this contribution we concentrate on how the public perceives the EU's role in climate governance in the EU's multilevel system.

Is there cross-country variation in the support for the EU's involvement in climate policy? Which characteristics of individuals make it more likely to support the EU's mandate in climate policy? Do citizens prefer to concentrate or to diffuse competence for climate policy in the EU's multilevel system? These three research questions guide our analysis in which we empirically address both country- and individual-level differences.

Our findings show that there is cross-country variation among the respondents from the individual EU countries. With regard to our second research question, knowledge on how the EU functions as well as the overall positive attitude towards the EU have a significant positive effect on respondents' support for the EU's involvement in climate policy. Turning to the third research question, support for climate policymaking at local and national levels is significantly and positively correlated with the willingness to support the EU level of decision-making. This finding suggests that Europeans are willing to diffuse the competence for climate policy across the individual levels of the EU's multilevel system. The broader implication of this finding is that all political levels have a mandate to produce climate policy innovations and to implement them. We believe that this can be interpreted as the individual levels should be more ambitious and active in addressing climate change by adopting appropriate policy measures. Furthermore, our findings suggest that the individual levels should consider cooperation more closely in their efforts to address climate change.

The remainder of this chapter unfolds as follows. First, we give an overview of the EU's legislative activities in the field of climate policy. Next, we detail our conceptual model and formulate empirically testable hypotheses. While we empirically also offer a descriptive analysis at the level of the EU member states, our theoretical considerations are constrained to the individual level. The subsequent section details the data utilized in this analysis and gives information on the analytical method. In the next section, we present results of our analysis and discuss the main findings. The presentation of the empirical findings includes both a descriptive analysis at the country level and a theory-testing analysis with individual-level data. The chapter ends with a summary of our findings followed by concluding remarks.

2   The EU's Legislative Activities Regarding Climate Change

Despite its complex institutional design, the European Union has been active and prolific in producing climate change legislation (Damro, Hardie, & MacKenzie, 2008; Rayner & Jordan, 2013). Climate change has been, in comparison with other environmental fields, particularly high on the EU agenda. Table 1 gives an overview of EU legislation on climate change adopted between 2000 and 2017.

The two most prominent mechanisms for responding to climate change are mitigation and adaptation. Climate mitigation policies

Table 1 Overview of EU climate legislation and relevant supporting documents

Year	Legislative document	Name
2018	Regulation Proposal COM/2016/0479	Regulation on the inclusion of greenhouse gas emissions and removals from land use, land use change and forestry (LULUCF)
2017	Decision 2017/175	Decision on establishing EU Ecolabel criteria for tourist accommodation
2014	Communication COM/2014/15	2030 framework for climate and energy policies
2014	Communication COM/2014/330	European Energy Security Strategy
2014	Regulation (EU) No. 517/2014	Regulation on fluorinated gases and repealing Regulation (EC) No. 842/2006
2014	Directive 2014/94/EU	Directive on the deployment of alternative fuels infrastructure
2013	Communication COM/2013/2016	EU Strategy on adaptation to climate change
2013	Decision 529/2013/EU	Decision on accounting rules on greenhouse gas emissions and removals resulting from activities relating to land use, land-use change and forestry and on information concerning actions relating to those activities
2012	Directive 2012/27/EU	Directive on energy efficiency, amending Directives 2009/125/EC and 2010/30/EU and repealing Directives 2004/8/EC and 2006/32/EC
2011	Regulation (EU) No. 510/2011	Regulation on setting of emission performance standard for new light commercial vehicles as part of the Union's integrated approach to reduce CO_2 emission from light-duty vehicles
2010	Communication SEC/2010/1346	Energy 2020: A strategy for competitive, sustainable and secure energy
2010	Directive 2010/30/EU	Directive on the indication by labelling and standard product information of the consumption of energy and other resources by energy-related products

(continued)

Table 1 (continued)

Year	Legislative document	Name
2009	2020 Climate & Energy Package	Common label for a set of policies: Directive 2009/29/EC, Directive 2009/28/EC, Directive 2009/31/EC and Decision No. 406/2009/EC. Goal of the package is to reduce EU GHG emissions by at least 20% below 1990 levels, to have 20% of renewables in EU energy consumption, and to attain 20% reduction in primary energy use, by improving energy efficiency
2009	Directive 2009/33/EC	Directive on promotion of clean and energy-efficient road transport vehicles
2009	Directive 2009/125/EC	Directive on establishment of a framework for the setting of eco-design requirements for energy-related products
2009	Directive 2009/72/EC	Directive concerning common rules for the internal market in electricity and repealing Directive 2003/54/EC
2009	Decision 406/2009/EC	Decision on the effort of Member States to reduce their greenhouse gas emissions to meet the Community's greenhouse gas emission reduction commitments up to 2020
2009	Regulation (EC) No. 443/2009	Regulation on setting of emission performance standards for new passenger cars as part of the Community's integrated approach to reduce CO_2 emissions from light-duty vehicles
2009	Directive 2009/30/EC	Directive on amendment of the Directive 98/70/EC as regards the specification of petrol, diesel and gas-oil and introducing a mechanism to monitor and reduce greenhouse gas emissions and amendment of the Council Directive 1999/32/EC as regards the specification of fuel used by inland waterway vessels and repealing Directive 93/12/EEC

(continued)

Table 1 (continued)

Year	Legislative document	Name
2009	Directive 2009/31/EC	Directive on geological storage of carbon dioxide
2009	Directive 2009/29/EC	Directive on amendment of the Directive 2003/87/EC on emission allowance trading scheme
2008	Regulation (EC) No. 71/2007	Regulation on setting up the Clean Sky Joint Undertaking
2004	Directive 2004/8/EC	Directive on promotion of cogeneration based on a useful heat demand in the internal energy market and amending Directive 92/42/EEC
2004	Directive 2003/96/EC	Directive on restructuration of the Community framework for the taxation of energy products and electricity
2004	Decision 280/2004/EC	Decision on accounting rules on greenhouse gas emissions and removals resulting from activities relating to land use, land-use change and forestry and on information concerning actions relating to those activities
2003	Directive 2003/87/EC	Directive on establishing a scheme for greenhouse gas emission allowance trading within the Community and amending Council Directive 96/61/EC
2002	Directive 2002/91/EC	Directive on energy performance of buildings
2000	Directive 1999/94/EC	Directive on Information on the fuel consumption and CO_2 emissions of new cars

Note Adapted from Fleig, Schmidt, and Tosun (2017)

comprise measures that aim at reducing or eliminating the drivers of climate change, while adaptation measures aim at managing or limiting the degree of actual or expected climate change impacts (Fleig et al., 2017).

At the EU level, several mitigation measures were adopted, which include an increased use of renewable energy and combined heat and power installations, improved energy efficiency in buildings, industry,

household appliances, the reduction of CO_2 emissions from new pas-
senger cars, abatement measures in the manufacturing industry, and
measures to reduce emissions from landfills (Tomescu et al., 2016).

The EU's landmark legislation, the Climate and Energy Package
which was adopted in 2009, pursues a triple goal: a 20% reduction of
Greenhouse Gas (GHG) emissions compared with 1990, a 20% share
of renewables in the EU energy consumption, and energy improvement
by 20% (Tomescu et al., 2016). To attain these goals, the EU adopted
an Emissions Trading Scheme (ETS) and an Effort Sharing Decision
for non-ETS target sectors, which covers CO_2 emissions in buildings,
transport and non-ETS industry and non-CO_2 greenhouse gas emissions (Harmsen, Eichhammer, & Wesselink, 2011). The EU ETS
mostly covers greenhouse gas emissions from large-scale facilities in the
power and industry sectors, as well as the aviation sector (Dessler &
Parson, 2010). The EU's mitigation efforts are supported by the energy
efficiency action plan, which will help to achieve a 20% energy saving
by 2020 (da Graça Carvalho, 2012). Overall, the EU's climate policy is
characterized by a multifaceted character, with the EU ETS presenting
the 'centerpiece' (Boasson, 2015, p. 54).

Adaptation measures concentrate on preparing for the effects of climate change in such a way that the expected damage will be prevented
or at least minimized (e.g. Brouwer, Rayner, & Huitema, 2013; Dupuis
& Biesbroek, 2013; Howlett, 2009; Nalau, Preston, & Maloney, 2015).
Most rampant adaptation measures concentrate on developing strategies
for using scarce water resources efficiently and managing drought periods,
adapting building to global temperature increases and extreme weather
conditions, and flood management. Policies can either be designed in
anticipation of change (anticipatory adaptation) or in response to changes
(reactive adaptation aimed at building resilience) (Adger, Arnell, &
Tompkins, 2005; Tubiana, Gemenne, & Magnan, 2010).

In April 2013, the European Commission adopted an EU strategy on
adaptation to climate change, which is markedly different from the mitigation measures promoted by the EU as it mostly encourages the member
states to adopt comprehensive adaptation strategies by means of knowledge
production and sharing and providing financial assistance. Furthermore,
the EU Commission supports the cities' roles in adapting to climate

change through the Mayors Adapt Initiative, a voluntary commitment within the framework of the Covenant of Mayors (Kona et al., 2016).

We expect that with growing legislative activities and corresponding news media coverage of these, citizens will become increasingly aware of the EU as an actor in these policy areas. Already feeling the effects of climate change in terms of extreme weather events or heavy rainfall may also increase their interest in this topic. Once becoming aware of the EU's role, the citizens will form an opinion on whether they are supportive of the EU's activities or not. A positive or negative public opinion on the EU's influence depends on a number of variables to which we will turn in the next section. The point we wanted to make in this section is that our notion of the EU as a relevant actor in climate politics is substantiated by its legislative activities. Of course, this does not preclude that other actors are equally or more relevant in the EU's multilevel system. To be sure, empirical research has shown that the local level is particularly active with regard to implementing climate measures (Damsø, Kjær, & Christensen, 2016) and individual EU countries have demonstrated their will to exceed the EU's level of ambition (Tobin, 2017), even if it is just for a fixed period of time (e.g. the UK; see Gillard & Lock, 2017). The coexistence of actors does not contradict our reasoning, but only suggests that we have to control for additional political levels when seeking to explain support for the EU's role in addressing climate change.

3 Conceptual Framework and Hypotheses

In this chapter, we are concerned with the European citizens' willingness and determining factors to delegate legal competences for governing climate change to the EU level. Usually, delegation is discussed jointly with accountability, as they form two directions of the same process. Accountability refers to relationships within democratic systems between citizens, who delegate decision-making power, and political actors, who receive it. The main mechanism through which political actors are accountable to citizens is through elections since during electoral campaigns political actors have to explain how they have addressed certain policy problems. If political actors do not live up to the citizens' expectations, they are voted out of office. Thus, in democratic systems, political actors have to justify and explain their policy action vis-à-vis citizens (Bäckstrand, 2006, p. 295).

Yet there exist different types of actors in democratic systems who are responsible for policymaking of which only some, the political actors, participate in elections and are directly accountable, and others, the bureaucratic actors, do not participate and are indirectly accountable. Furthermore, delegation—such as the shifting of competences—among political actors as well as between political and bureaucratic actors further affects the functioning of the electoral mechanism for accountability (Strøm, Bergman, & Müller, 2003). The delegation chain is even longer and therefore accountability further complicated in the EU's multilevel system (Bache, 2012; Bergman, 2000; Bovens, 2007; Herranz-Surrallés, 2017; Papadopoulos, 2010). At the same time, while delegation can obfuscate accountability, it can help to increase a system's problem-solving capacity (Scharpf, 1997). Therefore, delegation is 'Janus-headed' and research on this topic has to acknowledge this feature (Maggetti & Trein, 2019).

Our conceptual framework concentrates on delegation only and differentiates between three sets of variables. The first set of variables takes into account the multilevel system of the EU and elaborates on whether Europeans are likely to distribute competence across the political levels or concentrate it.

Political systems are socially constructed entities comprising both formalized structures (such as rules) and informalized codes and expectations by their participants (Craig, 2015). The process that leads to shared understandings of formal and informal constructions in political economies is socialization (Mendoza, 2007). We argue that the socialization processes in political systems are strong enough to affect the individuals' attitudes on the allocation of policy competences. This argument aligns with newer contributions to the literature on political economy that shows how macro-level phenomena affect the individual's perceptions (Walter, 2017). Consequently, the more individuals are socialized with the existence of a European approach to a policy, the more they will support the delegation of competences to the EU level.

Climate change is a collective action problem of immense complexity and spans across national borders and permeates into many spheres of life (Jordan, Huitema, van Asselt, 2011; Ostrom, 2010; Sovacool, 2011). What is also important is that the European Commission has been quick in including climate policy in its political agenda and therefore the EU level plays a considerable role in European climate policy (Biedenkopf, 2014; Boasson & Wettestad, 2016; Oberthür & Dupont, 2015; Tosun & Solorio, 2011; Schmidt & Fleig, 2018). Given the level of policy activity triggered by the EU Commission, we contend that socialization arising from this policy arrangement leads to a situation in which European citizens are willing to spread policy competences across the various levels of the political system. This reasoning culminates in our first two hypotheses in which we postulate a positive relationship between the willingness to delegate authority to the EU and to the national/regional level.

H1: Support for climate policy at the local/regional level increases support for climate policy at the EU level.

H2: Support for climate policy at the national level increases support for climate policy at the EU level.

The second set of hypotheses concern the individuals' knowledge about and attitudes towards the EU. We posit that a better understanding of how the EU functions results in a greater willingness to delegate policy competences regarding climate change to that level. This argument is in line with previous research on links between knowledge about the EU and the sense of belonging to the European community. The assumption is that the more citizens know about the EU, the more they will identify with it. Consequently, they will feel more related to this level of policymaking (Faas, 2007; Ingelhart, 1970; Thorpe, 2008). The second argument in favour of our hypothesis comes from the theory on economic utilitarianism and the EU membership. According to this view, citizens are more in favour of the EU if they expect positive outcomes from the membership of their country in the EU (Cinnirella, 1997; Fligstein, Polyakova, & Sandholtz, 2012; Kritizinger, 2005). In line with this argument, individuals who are concerned about climate change should understand that tackling climate change is more effective when done at the European level. Furthermore, empirical research has shown that competitive considerations induce the EU to 'export' its climate policy to other jurisdictions, which leads to a further strengthening of climate protection at the international level (Jänicke & Quitzow, 2017; Schreurs & Tiberghien, 2010). As a result, we expect individuals with knowledge about the EU to be more receptive of the idea of delegating policymaking competence to that political level.

H3: Knowledge on how the EU functions increases support for climate policy at the EU level.

The fourth hypothesis draws on the reasoning that individuals will be more likely to delegate political power to the EU when they hold a positive attitude about European integration. Positive attitude is in essence an emotional or 'gut commitment' (Boomgaarden, Schuck, Elenbaas, & de Vreese, 2011, p. 243) to a specific political object, such as a state, the EU, or both. The aforementioned reasoning should not only hold true for climate policy, but also for any policy area that falls under the competence of the EU. By the same token, individuals who hold a critical view of the EU are likely to abstain from delegating the EU any competences. Here, we follow the positive connotation of our reasoning and put forward the hypothesis accordingly.

H4: Positive attitudes about the EU increase support for climate policy at the EU level.

We also argue that perceiving climate change as a serious threat will induce respondents to assign policymaking competences to the EU, probably along with other public entities such as national and regional governments and cities. This assumption is in line with findings from the literature that risk perception is a significant predictor of willingness to support governmental action to tackle climate change (O'Connor, Bord, & Fisher, 1999).

H5: A perception of climate change as a serious threat increases support for climate policy at the EU level.

Our last hypothesis is based on the literature on political trust in the EU institutions and variations among EU member states. An analysis by Arnold, Sapir, and Zapryanova (2012) shows that trust in the EU institutions correlates with trust in the national political institutions. Moreover, there is a clear difference between old member states and new member states in levels of trust, with citizens of old member states being as much as twice more likely to trust party system, parliament and judiciary system (Arnold et al., 2012). Having said that, we expect that citizens from countries with lower levels of trust in the EU institutions—new member states—will also show lower levels of support for the EU level of decision-making. Therefore, we propose the hypothesis accordingly.

H6: Citizens of the new EU member states have lower levels of support for the EU policymaking than citizens of the old member states.

4   Clarifications on Data and Measurement

The main challenge of testing the hypotheses formulated above is the availability and access to adequate data. The best data source available for addressing these research questions are the Eurobarometer surveys. Standard surveys are repeated periodically, while special opinion surveys are conducted occasionally. Survey design and execution correspond to guidelines and rules in social science research (Leong & Ward, 2006). The surveys are carried out by polling firms based in the individual member states. All interviews were conducted face-to-face, in households, using the national language of the respective state. The number of respondents in each state, except Malta, Cyprus and Luxembourg, is at least 1,000 with all of them being over the age of 15. For the purpose of this analysis, we rely on the Eurobarometer 87.1 issue published in 2017 (European Commission & European Parliament, 2017).

The question of interest for this analysis is: 'In your opinion, who within the EU is responsible for tackling climate change?' (QC3). The response categories for this question are greater in number including: 'national government', 'European Union', 'business and industry, 'regional and local authorities', 'you personally', 'environmental groups', 'other', 'all of them', 'none', and 'don't know'. However, for the present analysis, we only take into account the 'European Union' as the response category that operationalizes our dependent variable. We take into account two more response categories as measurements of the first two explanatory variables, namely *Responsibility National* and *Responsibility Regional*. The first variable gauges which of the respondents indicated that in the EU national governments should be responsible for tackling climate change and the second whether the responsibility was assigned to regional and local authorities. For the purpose of this analysis, we excluded responses that concern non-state actors since assigning responsibility to them would entail the need to venture into a different literature strand, that is the literature on public–private partnerships (Pattberg, 2010) or transnational governance (Andonova et al., 2014). This is insightful literature, but we could not do justice to it here and therefore our analysis is constrained to the assignment of competences to public actors only. The dependent variable and the two explanatory variables are dichotomous.

The next set of explanatory variables assessed the surveyed individuals' stance on the EU. The variable *Knowledge EU* is a particularly interesting measurement that counts the number of correct responses to three questions on the functioning of the EU. The first of these questions concerns the statement that the European Parliament elects the president of the European Commission, the second that the European Parliament participates jointly with the member states in EU policymaking, and the third that each state has the same number of members in the European Parliament. *Knowledge EU* ranges from 0 to 3, depending on how many of these questions were answered correctly by the respondents. *Attitude EU* gauges whether the respondents regard EU membership as a 'good thing' (3), 'neither a good thing nor a bad thing' (2), or 'bad thing' (1). As hypothesized above, we believe it is reasonable to expect that respondents who generally support the EU are also more likely to assign the responsibility for climate policy to the EU.

We continue with the variables that assess the perception of the respondents concerning the seriousness of climate change. To this end, we rely on two measurements. The first one, *Seriousness*, indicates how the respondents rate climate change on a 10-point scale ranging from 'not at all a serious problem' (1) to 'an extremely serious problem' (10). We complement this measurement by a second one called *Most serious*, which indicates which respondents mentioned climate change as the world's most serious problem. While this binary variable represents a particularly strict measurement of the respondents' perception, it is surprising that of all surveyed individuals about one-third agreed with this statements.

We now turn to the battery of control variables with *Internet use* being the first. This variable indicates the frequency of internet use ranging from 1 (never) to 7 (very frequently). We include internet use since studies have shown that it has an impact on the formation of attitudes towards climate change, albeit the effect is a complex one (Taddicken, 2013). Furthermore, we control for the respondents' gender since studies have shown that women tend to express greater concern about climate change than do men (McCright, 2010). Another widely used control variable when testing the determinants of attitudes towards climate change is the respondents' age. It is generally expected that younger people are more aware of climate change and express greater concern (Poortinga, Spence, Whitmarsh, Capstick, & Pidgeon, 2011).

Lastly, we control for the occupation of the respondents, which is motivated by two considerations. First, the overall support for the EU is affected by whether people think that they are benefitting from the EU. Research has shown that 'white collar' respondents are more likely to support European integration, but more recently studies have alluded that those who are economically most vulnerable—the unemployed—see European integration as a strategy that might improve their situation (Kuhn & Stoeckel, 2014). Second, the occupation group also affects people's attitudes towards climate change as shown, for example, by Pietsch and McAllister (2010). Therefore, we differentiate between individuals who are self-employed, managers, white-collar workers, manual workers, house persons, unemployed and students. Table 2 presents the descriptive statistics of the variables used for the analysis.

Since the dependent variable of this study is binary, we fit multilevel logit models with random intercepts. Given our theoretical interest, we decided against including state-level variables, but to run the analyses

Table 2 Descriptive statistics

Variable	Obs.	Mean	Std. Dev.	Min.	Max.
Responsibility EU	26,318	0.4027282	0.4904562	0	1
Focal explanatory variables—Multilevel governance					
Responsibility National	26,318	0.4574056	0.4981919	0	1
Responsibility Regional	26,318	0.2458014	0.4305695	0	1
Focal explanatory variables—EU					
Knowledge EU	26,318	1.888327	0.9975231	0	3
Attitude EU	26,318	2.42412	0.7257621	1	3
Focal explanatory variables—Problem perception					
Seriousness	26,318	7.593396	2.113401	1	10
Most serious	26,078	0.3045095	0.4602082	0	1
Control variables					
Internet use	26,318	5.672088	2.121532	1	7
Female	26,318	0.547268	0.4977702	0	1
Age	26,318	51.4343	18.05993	15	99
Self-employed	26,318	0.0638726	0.2445306	0	1
Manager	26,318	0.1110647	0.3142182	0	1
Other white collar	26,318	0.1175621	0.322095	0	1
Manual worker	26,318	0.1912759	0.3933133	0	1
House person	26,318	0.0574892	0.2327793	0	1
Unemployed	26,318	0.0647846	0.24615	0	1
Retired	26,318	0.3347899	0.4719259	0	1
Student	26,318	0.059161	0.2359303	0	1

Note All data taken from European Commission and European Parliament (2017)

with individual-level variables only. From this perspective, the estimation models take into account the intercepts that vary across the individual states. We will discuss country-level factors when reporting the descriptive findings for the mean concern expressed by the respondents in the individual member countries. It should be noted that the number of cases varies between the descriptive and the multilevel analysis. In the latter, we rely on 30 cases since Eurobarometer differentiates between East and West Germany as well as Great Britain and Northern Ireland, whereas in the first data for the 28 member states the data for Germany and Great Britain are merged.

5   Presentation and Discussion of the Findings

In this section, we present the findings of our empirical analysis. We begin by offering a descriptive analysis of the variation across countries. In the next step, we turn to the findings for the individual level. While we observe differences across countries, we can identify individual-level characteristics that make it more likely for European citizens to delegate the political competence for governing climate change to the EU.

5.1   Analysis at the Country Level

Figure 1 shows the cross-country variation in the mean value regarding the agreement on the statement that the EU is responsible for tackling climate change. Average aggregated support on the basis of 27,901 respondents is above 39%. The values of the variable on the country level range from 23 to 68% (see Fig. 1).

Three countries, namely Estonia and Latvia (both 23%) and the UK with 25%, obtained the lowest percentage of agreement to the aforementioned statement. Taking out Northern Ireland and only looking at Great Britain shows an even lower percentage rate of 22%. This is a surprising finding considering the fact that the UK has played a leading role in supporting EU climate policies. Particularly during the Rio and Kyoto summits, the UK has been characterized by exhibiting international leadership and, for instance, reached the EU's Kyoto Protocol emission reduction target prior to schedule (Jordan & Rayner, 2016). In 2008, the UK took strong action at the national level by adopting the Climate Change Act. This innovative and highly determined piece of legislation targeted the reduction of emissions by 80% from 1990 levels by 2050 (Lorenzoni & Benson, 2014). However, the failure to deliver an agreement at the negotiations in Copenhagen in 2009, and especially the financial crisis, lowered efforts to achieve a low carbon transition in the UK (Burns & Tobin, 2016; Gillard & Lock, 2017). In addition, the timing of the opinion survey coincided with the

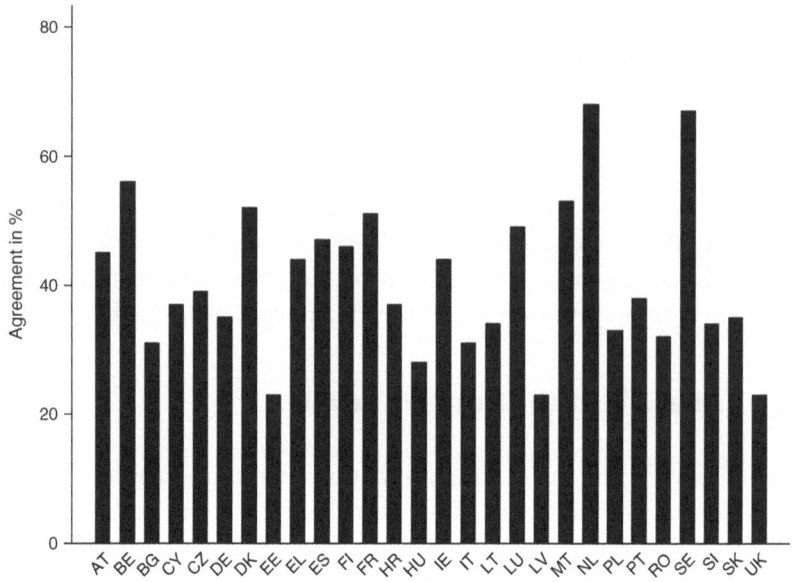

Fig. 1 Agreement to EU's involvement in climate policy, 2017 (*Note* Own illustration based on Eurobarometer [2017])

period in which the UK voted to leave the EU. This circumstance may also explain these low levels of agreement.

A number of states showed levels around 30% agreement, namely Hungary (28%), Italy (31%), Bulgaria (31%), Poland (32%), Romania (33%), Slovenia (33%), Czech Republic (33%), Slovakia (35%) and Germany (35%). Slightly higher agreements of nearly and over 40% are observable in Cyprus (37%), Portugal (38%), Austria (41%), Ireland (43%), Greece (44%), Croatia (46%), Spain (47%), Finland (47%) and Malta (49%). Interestingly, half of the respondents in France (50%) and Luxembourg (51%) supported the statement. The highest levels of approval regarding the EU's responsibility to tackle climate change come from Belgium (56%), Sweden (63%) and the Netherlands (67%). In addition, it is important to note that for the state scoring the highest value, the Netherlands, climate change is a tremendously important issue. As parts of the country lie below sea level, sea level rise has a peculiar meaning. However, Pettenger (2016) is eager to note that efforts by the Dutch are not only driven by vulnerability, but also stem from an understanding to promote sustainable development and climate action domestically and at the EU level.

Excluding Croatia and Malta, what is interesting to notice here is that states showing above-average support for the statement are old member states of the EU. On the one hand, there is a clear trend that new member states have below-average support for the statement and the largest part of them has values below 35%. For instance, we know from the case of Poland that climate change is a topic which does not overly concern the public or is present in parliamentary debates (Marcinkiewicz & Tosun, 2015).

On the other hand, out of the old member states, Italy, Germany and the UK show below-average support for the statement that the EU is responsible for tackling climate change. However, particularly Germany's case in point is a surprising finding. Despite being one of the leading countries in Europe to develop and implement climate policies as well as a strong supporter of EU legislation (Weidner & Mez, 2008), it shows below-average support for the statement. It might be the case that German public discourse on climate change is so heavily influenced by the national energy transition plan (*Energiewende*), that a large part of the public identifies the national and below-national levels as more important ones. Moreover, a possible explanation could be that public views are such, that an ambitious climate action like *Energiewende,* due to its scale and especially costs, is considered an unattainable goal for the rest of the EU. Thus, shifting decision-making capabilities to the EU level would indubitably cause decreased policy ambitions at the national level. However, more research is needed to support these claims. Findings from the state analysis, despite three exceptions, support our expectations formulated in H6.

5.2   Analysis at the Individual Level

Table 3 presents three multilevel logit models. The first model presents the covariates associated with Hypotheses 1 through 4. The second model includes the indicators associated with Hypothesis 5 together with the

Table 3 Multilevel model

	Model 1	Model 2	Model 3
Responsibility National	1.073 (0.029)***		1.051 (0.029)***
Responsibility Regional	1.106 (0.034)***		1.083 (0.034)***
Knowledge EU	0.105 (0.015)***		0.090 (0.015)***
Attitude EU	0.183 (0.021)***		0.157 (0.021)***
Seriousness		0.100 (0.007)***	0.065 (0.007)***
Most serious		0.159 (0.029)***	0.090 (0.031)***
Internet use		0.023 (0.008)***	−0.000 (0.009)
Female		−0.163 (0.027)***	−0.142 (0.029)***
Age		−0.001 (0.001)	−0.003 (0.001)**
Self-employed		Base category –	Base category –
Manager		0.106 (0.065)	0.043 (0.070)
Other white collar		0.039 (0.064)	0.021 (0.069)
Manual worker		−0.060 (0.060)	−0.005 (0.065)
House person		−0.097 (0.078)	0.001 (0.084)
Unemployed		−0.054 (0.074)	0.051 (0.079)
Retired		−0.067 (0.063)	−0.003 (0.067)
Student		−0.024 (0.081)	0.002 (0.088)
Intercept	−1.860 (0.092)***	−1.186 (0.141)***	−2.030 (0.150)***
Intercept country level	0.154 (0.041)***	0.183 (0.049)***	0.147 (0.040)***
Akaike Information Criterion	30,333.151	33,628.605	30,042.433
Observations	26,318	26,078	26,078
Cases	30	30	30

*$p < 0.10$; **$p < 0.05$; ***$p < 0.01$

control variables. The third model includes all variables. This sequential estimation technique allows us to vary the test conditions for the variables and to check the robustness of these findings. The tables report the logit coefficients, which we can only interpret in terms of the direction of the effects and whether the effects are significantly different from zero.

When observing the findings for Model 1, we see that all four variables produce positive coefficients that are significant at the 1%-level. This means that respondents who agree that the national and the regional levels need to be assigned competences for addressing climate change have a greater chance of agreeing that the EU should receive these competences. With increasing knowledge about how the EU functions, the chances also increase that the respondents assign to it competencies for climate policy. Lastly, a more positive attitude on the EU also increases the chances of allocating policy competence for addressing climate change to the EU. Based on Model 1, we can confirm our first four hypotheses.

Turning to Model 2, we can also confirm Hypothesis 5 and what is remarkable is that both operationalizations of the theoretical construct produce positive and significant coefficients. It should be noted that the two variables correlate only weakly with each other. Of the control variables, internet use is also positively and significantly correlated with the likelihood of a respondent allocating political competence to the EU. By contrast, the variable gauging gender has the opposite effect: according to Model 2, being a woman significantly reduces the likelihood of assigning policymaking competence to the EU. The other control variables fail to produce coefficients that are significantly different from zero.

Model 3 presents the full model and confirms the findings of the previous two models. The five focal explanatory variables produce positive and significant coefficients, supporting the logic underlying our five hypotheses. From this, it follows that European citizens prefer diffusing political competence for addressing climate change across different levels. Our findings also show that knowledge about and a positive attitude on the EU matter for the individuals' willingness to delegate competence to the European level. Finally, perceiving climate change to be the most serious issue is conducive to delegating competence to the EU.

In sum, these findings allude to an opportunity structure for the EU to take more ambitious action concerning the governance of climate change. Our results also speak to the EU's self-perception of being a leader in international climate governance. The citizens in the EU appear to be willing to accept that the EU takes the role as a leader and pushes for a more effective climate regime. In other words, the EU has the domestic legitimacy to perform such as role, and we would go one step further and argue that this finding entails the responsibility to act more firmly on climate governance.

6   Conclusion

When thinking about the causes of climate change and the attempts to mitigate it, it becomes clear why scholarship has embraced the governance concept (Pattberg, 2010; Tosun & Schoenefeld, 2017). Greenhouse gas emissions are produced by a multitude of actors who are based in different action contexts, and reducing these emissions, in an attempt to limit global warming to below 2 °C, will critically depend on the formation of multi-stakeholder partnerships.

In this chapter, we observed that respondents are willing to delegate climate policy competence to the European level, although the willingness varies across the individual member states. When looking at which individual-level factors matter for the willingness to assign the EU the competence to make climate policy, knowledge about how the EU functions and a generally positive attitude about the EU increase the likelihood for such a response. Respondents that regard climate change to be a serious or even the most serious threat are also more likely to delegate the policymaking competence to the EU. A third finding relates to whether the respondents assigning competence to the regional and/or national level are either more or less likely to delegate decision-making competence to the EU. Our findings clearly show that individuals who assign this competence to the regional and/or national level are also more likely to attribute this power to the EU. We interpret this finding as one that respondents are willing to diffuse political power across different levels of the EU's multilevel system. They do not regard the individual levels as competing but complementary, and this is an adequate perception of the factual policymaking situation. We invite future research to look into this finding and to reflect on it by applying the perspective of polycentric climate governance and policy experiments at the local level (see, e.g., Jordan et al., 2015; Ostrom, 2010; Sovacool, 2011). The concept of polycentric governance explicitly posits that different governance modes must be used simultaneously rather than in isolation (Jordan et al., 2015).

Overall, we believe that concentrating on the perceptions of EU citizens on delegation of policy competences in climate policy is a promising perspective for advancing the state of knowledge and for being in a better position to inform policymaking. If the EU wants to attain further policy integration in the domain of climate policy, it is well advised to take public opinion on climate issues into account. Climate policymaking competencies need to be highlighted to the citizenship if the EU seeks to increase accountability and thereby improve the quality of democratic decision-making. We believe that bringing in considerations concerning accountability and delegation has the potential to constitute a new and exciting research perspective as well as to inform a more precautionary approach to governing climate change (see Tosun, 2013). Consequently, we propose it as a prominent perspective for future research endeavours.

References

Adger, W. N., Arnell, N. W., & Tompkins, E. L. (2005). Successful adaptation to climate change across scales. *Global Environmental Change, 15*(2), 77–86.

Andonova, L., Betsill, M. M., Bulkeley, H., Compagnon, D., Hale, T., Hoffmann, M. J., ..., Van Deveer, S. D. (2014). *Transnational climate change governance*. Cambridge: Cambridge University Press.

Arnold, C., Sapir, E., & Zapryanova, G. (2012). Trust in the institutions of the European Union: A cross-country examination. *European Integration online Papers (EIoP), 16*, 1–39.

Bache, I. (2012). Multi-level governance in the European Union. In D. Levi-Faur (Ed.), *The Oxford handbook of governance*. Oxford: Oxford University Press.

Bäckstrand, K. (2006). Multi-stakeholder partnerships for sustainable development: Rethinking legitimacy, accountability and effectiveness. *Environmental Policy and Governance, 16*(5), 290–306.

Bergman, T. (2000). The European Union as the next step of delegation and accountability. *European Journal of Political Research, 37*(3), 415–429.

Biedenkopf, K. (2014). The European Parliament in EU external climate governance. In D. Irrera & S. Stavridis (Eds.), *The European Parliament and its international relations* (pp. 92–108). Abingdon: Routledge.

Boasson, E. L. (2015). *National climate policy: A multi-field approach*. Abingdon: Routledge.

Boasson, E. L., & Wettestad, J. (2016). *EU climate policy: Industry, policy interaction and external environment*. Abingdon: Routledge.

Boomgaarden, H. G., Schuck, A. R. T., Elenbaas, M., & de Vreese, C. (2011). Mapping EU attitudes: Conceptual and empirical dimension of Euroscepticism and EU support. *European Union Politics, 12*(2), 231–266.

Bovens, M. (2007). New forms of accountability and EU-governance. *Comparative European Politics, 5*(1), 104–120.

Brouwer, S., Rayner, T., & Huitema, D. (2013). Mainstreaming climate policy: The case of climate adaptation and the implementation of EU water policy. *Environment and Planning C: Government and Policy, 31*(1), 134–153.

Burns, C., & Tobin, P. (2016). The impact of the economic crisis on European Union environmental policy. *JCMS: Journal of Common Market Studies, 54*(6), 1485–1494.

Cinnirella, M. (1997). Towards a European identity? Interactions between the national and European social identities manifested by university students in Britain and Italy. *British Journal of Social Psychology, 36*(1), 19–31.

Craig, M. (2015). Post-2008 British industrial policy and constructivist political economy: New directions and new tensions. *New Political Economy, 20*(1), 107–125.

da Graça Carvalho, M. (2012). EU energy and climate change strategy. *Energy, 40*(1), 19–22.

Damro, C., Hardie, I., & MacKenzie, D. (2008). The EU and climate change policy: Law, politics and prominence at different levels. *Journal of Contemporary European Research, 4*(3), 179–192.

Damsø, T., Kjær, T., & Christensen, T. B. (2016). Local climate action plans in climate change mitigation—Examining the case of Denmark. *Energy Policy, 89,* 74–83.

Dessler, A., & Parson, E. (2010). *The science and politics of global climate change: A guide to the debate.* Cambridge: Cambridge University Press.

Dupuis, J., & Biesbroek, R. (2013). Comparing apples and oranges: The dependent variable problem in comparing and evaluating climate change adaptation policies. *Global Environmental Change, 23*(6), 1476–1487.

Egeberg, M., & Trondal, J. (2016). Why strong coordination at one level of government is incompatible with strong coordination across levels (and how to live with it): The case of the European Union. *Public Administration, 94*(3), 579–592.

European Commission and European Parliament. (2017). Eurobarometer 87.1 (Version 1.2.0) [Study No. ZA6861]. Cologne, GESIS: Leibniz Institute for the Social Sciences. https://doi.org/10.4232/1.12922.

Faas, D. (2007). Youth, Europe and the nation: The political knowledge, interests and identities of the new generation of European youth. *Journal of Youth Studies, 10*(2), 161–181.

Fleig, A., Schmidt, N. M., & Tosun, J. (2017). Legislative dynamics of mitigation and adaptation framework policies in the EU. *European Policy Analysis, 3*(1), 101–124.

Fligstein, N., Polyakova, A., & Sandholtz, W. (2012). European integration, nationalism and European identity. *Journal of Common Market Studies, 50*(1), 106–122.

Gillard, R., & Lock, K. (2017). Blowing policy bubbles: Rethinking emissions targets and low-carbon energy policies in the UK. *Journal of Environmental Policy & Planning, 19*(6), 638–653.

Harmsen, R., Eichhammer, W., & Wesselink, B. (2011). Imbalance in Europe's effort sharing decision: Scope for strengthening incentives for energy savings in the Non-ETS sectors. *Energy Policy, 39*(10), 6636–6649.

Herranz-Surrallés, A. (2017). Energy diplomacy under scrutiny: Parliamentary control of intergovernmental agreements with third-country suppliers. *West European Politics, 40*(1), 183–201.

Howlett, M. (2009). Governance modes, policy regimes and operational plans: A multi-level nested model of policy instrument choice and policy design. *Policy Sciences, 42*(1), 73–89.

Huitema, D., Jordan, A., Massey, E., Rayner, T., van Asselt, H., Haug, C., …, Stripple, J. (2011). The evaluation of climate policy: Theory and emerging practice in Europe. *Policy Sciences, 44*, 179–198.

Inglehart, R. (1970). Cognitive mobilization and European identity. *Comparative Politics, 3*(1), 45–70.

Jänicke, M., & Quitzow, R. (2017). Multi-level reinforcement in European Climate and Energy Governance: Mobilizing economic interests at the sub-national levels. *Environmental Policy and Governance, 27*(2), 122–136.

Jordan, A., & Rayner, T. (2016). The United Kingdom: A record of leadership under threat. In *The European Union in international climate change politics* (pp. 197–212). London: Routledge.

Jordan, A., Huitema, D., & van Asselt, H. (2011). Climate change policy in the European Union: An introduction. In A. Jordan, D. Huitema, H. Van Asselt, T. Rayner, & F. Berkhout (Eds.), *Climate change policy in the European Union*. Cambridge: Cambridge University Press.

Jordan, A. J., Huitema, D., Hildén, M., van Asselt, H., Rayner, T. J., Schoenefeld, J. J., …, Boasson, E. L. (2015). Emergence of polycentric climate governance and its future prospects. *Nature Climate Change, 5*(11), 977–982.

Jordan, A., Wurzel, R., Zito, A. R., & Brückner, L. (2003). European governance and the transfer of 'new' environmental policy instruments (NEPIs) in the European Union. *Public Administration, 81*(3), 555–574.

Keleman, R. D. (2010). Globalizing European Union environmental policy. *Journal of European Public Policy, 17*(3), 335–349.

Keleman, R. D., & Vogel, D. (2010). Trading places: The role of the United States and the European Union in international environmental politics. *Comparative Political Studies, 43*(4), 427–456.

Kona, A., Bertoldi, P., Melica, G., Rivas Calvete, S., Zancanella, P., Serrenho, T., …, Janssens-Maenhout, G. (2016). *Covenant of Mayors: Monitoring indicators* (Report No. EUR 27723 EN). Luxembourg: Publications Office of the European Union.

Kritzinger, Sylvia. (2005). European identity building from the perspective of efficiency. *Comparative European Politics, 3*(1), 50–75.

Kuhn, T., & Stoeckel, F. (2014). When European integration becomes costly: The euro crisis and public support for European economic governance. *Journal of European Public Policy, 21*(4), 624–641.

Leong, C. H., & Ward, C. (2006). Cultural values and attitudes toward immigrants and multiculturalism: The case of the Eurobarometer survey on racism and xenophobia. *International Journal of Intercultural Relations, 30*(6), 799–810.

Lorenzoni, I., & Benson, D. (2014). Radical institutional change in environmental governance: Explaining the origins of the UK Climate Change Act 2008 through discursive and streams perspectives. *Global Environmental Change, 29,* 10–21.

Maggetti, M., & Trein, P. (2019). Multi-level governance and problem-solving: Towards a dynamic theory of multi-level policymaking? *Public Administration,* forthcoming.

Marcinkiewicz, K., & Tosun, J. (2015). Contesting climate change: Mapping the political debate in Poland. *East European Politics, 31*(2), 187–207.

McCright, A. M. (2010). The effects of gender on climate change knowledge and concern in the American public. *Population and Environment, 32*(1), 66–87.

Mendoza, P. (2007). Academic capitalism and doctoral student socialization: A case study. *The Journal of Higher Education, 78*(1), 71–96.

Moravcsik, A. (2005). The European constitutional compromise and the neofunctionalist legacy. *Journal of European Public Policy, 12*(2), 349–386.

Nalau, J., Preston, B. L., & Maloney, M. C. (2015). Is adaptation a local responsibility? *Environmental Science & Policy, 48,* 89–98.

Oberthür, S., & Dupont, C. (Eds.). (2015). *Decarbonization in the European Union: Internal policies and external strategies.* London: Palgrave Macmillan.

Oberthür, S., & Gehring, T. (Eds.). (2006). *Institutional interaction in global environmental governance: Synergy and conflict among international and EU policies.* Cambridge: MIT Press.

Oberthür, S., & Roche Kelly, C. (2008). EU leadership in international climate policy: Achievements and challenges. *The International Spectator, 43*(3), 35–50.

O'Connor, R. E., Bord, R. J., & Fisher, A. (1999). Risk perceptions, general environmental beliefs, and willignes to address climate change. *Risk Analysis, 19*(3), 461–471.

Ostrom, E. (2010). Polycentric systems for coping with collective action and global environmental change. *Global Environmental Change, 20*(4), 550–557.

Papadopoulos, Y. (2010). Accountability and multi-level governance: More accountability, less democracy? *West European Politics, 33*(5), 1030–1049.

Pattberg, P. (2010). Public–private partnerships in global climate governance. *Wiley Interdisciplinary Reviews: Climate Change, 1*(2), 279–287.

Pettenger, M. E. (2016). The Netherlands' climate change policy: Constructing themselves/constructing climate change. In *The social construction of climate change* (pp. 75–98). Abingdon: Routledge.

Pietsch, J., & McAllister, I. (2010). 'A diabolical challenge': Public opinion and climate change policy in Australia. *Environmental Politics, 19*(2), 217–236.

Poortinga, W., Spence, A., Whitmarsh, L., Capstick, S., & Pidgeon, N. F. (2011). Uncertain climate: An investigation into public scepticism about anthropogenic climate change. *Global Environmental Change, 21*(3), 1015–1024.

Rayner, T., & Jordan, A. (2013). The European Union: The polycentric climate policy leader? *Wiley Interdisciplinary Reviews: Climate Change, 4*, 75–90.

Scharpf, F. W. (1997). Introduction: The problem-solving capacity of multi-level governance. *Journal of European Public Policy, 4*(4), 520–538.

Schmidt, N. M., & Fleig, A. (2018). Global patterns of national climate policies: Analyzing 171 country portfolios on climate policy integration. *Environmental Science & Policy, 84*, 177–185.

Schreurs, M., & Tiberghien, Y. (2010). Multi-level reinforcement: Explaining EU leadership in climate change mitigation. *Global Environmental Politics, 7*(4), 19–46.

Schulze, K., & Tosun, J. (2013). External dimensions of European environmental policy: An analysis of environmental treaty ratification by third states. *European Journal of Political Research, 52*(2), 581–607.

Schulze, K., & Tosun, J. (2016). Rival regulatory regimes in international environmental politics: The case of biosafety. *Public Administration, 29*(1), 57–72.

Selin, H., & Van Deveer, S. D. (2015). *European Union and environmental governance*. Abingdon: Routledge.

Sovacool, B. K. (2011). An international comparison of four polycentric approaches to climate and energy governance. *Energy Policy, 39*(6), 3832–3844.

Strøm, K., Bergman, T., & Müller, W. C. (Eds.). (2003). *Delegation and accountability in parliamentary democracies*. Oxford: Oxford University Press.

Taddicken, M. (2013). Climate change from the user's perspective: The impact of mass media and internet use and individual and moderating variables on knowledge and attitudes. *Journal of Media Psychology: Theories, Methods, and Applications, 25*(1), 39–52.

Thorpe, C. (2008). The distinguishing function of European identity: Attitudes towards and visions of Europe and the European Union among young Scottish adults. *Perspectives on European Politics and Society, 9*(4), 499–513.

Tobin, P. (2017). Leaders and laggards: Climate policy ambition in developed states. *Global Environmental Politics, 17*(4), 28–47.

Tomescu, M., Moorkens, I., Wetzels, W., Emele, L., Förster, H., & Greiner, B. (2016). *Renewable energy in Europe in 2016: Recent growth and knock-on effects.* Luxembourg: European Environmental Agency.

Tosun, J. (2013). How the EU handles uncertain risks: Understanding the role of the precautionary principle. *Journal of European Public Policy, 20*(10), 1517–1528.

Tosun, J., & Schoenefeld, J. J. (2017). Collective climate action and networked climate governance. *Wiley Interdisciplinary Reviews: Climate Change, 8*(1), 1–17.

Tosun, J. & Schulze, K. (2015). Compliance with EU biofuel targets in South-Eastern and Eastern Europe: Do interest groups matter? *Environment and Planning C: Government and Policy, 33*(5), 950–968.

Tosun, J., & Solorio, I. (2011). Exploring the energy–environment relationship in the EU: Perspectives and challenges for theorizing and empirical analysis. *European Integration online Papers, 15,* 1–11.

Tubiana, L., Gemenne, F., & Magnan, A. (2010). *Anticiper pour s'adapter. Le nouvel enjeu du changement climatique.* Paris: Pearson Education.

Vogel, D. (2003). The Hare and the Tortoise revisited: The new politics of consumer and environmental regulation in Europe. *British Journal of Political Science, 33*(4), 557–580.

Walter, S. (2017). Globalization and the demand-side of politics: How globalization shapes labour market risk perceptions and policy preferences. *Political Science Research and Methods, 5*(1), 55–80.

Weidner, H., & Mez, L. (2008). German climate change policy: A success story with some flaws. *The Journal of Environment & Development, 17*(4), 356–378.

7

Selected Studies
in Economic and Environmental
Accountabilities—Section Two

Beth Edmondson and Stuart Levy

Sustained analysis of selected climate change governance approaches affirms the importance of accountability for achieving both good environmental outcomes and enduring societies. These studies also suggest that institutional adaptiveness supports progressive implementation of international climate change adaptation policies across multiple decision-making levels. By actively supporting experimental approaches in overcoming stall-points or other limits to progress, adaptive institutions are well placed to maintain political goodwill with sustained policy buy-in, when obstacles or conflicts arise. The studies that follow suggest that climate governance mechanisms that make more use of accountability and monitoring opportunities are better able to avoid

B. Edmondson (✉)
School of Arts, Federation University, Churchill, VIC, Australia
e-mail: beth.edmondson@federation.edu.au

S. Levy
School of Education, Federation University, Churchill, VIC, Australia
e-mail: stuart.levy@federation.edu.au

© The Author(s) 2019
B. Edmondson and S. Levy (eds.), *Transformative Climates and Accountable Governance*, Palgrave Studies in Environmental Transformation, Transition and Accountability, https://doi.org/10.1007/978-3-319-97400-2_7

the limits of spiralling risk avoidance approaches and also more likely to pursue more flexible approaches that emphasise the benefits of change.

As shown in the following studies, effective climate governance relies upon a range of institutional features and accountable actors and agents. Key attributes include:

1. Scalable solutions
2. Internal and external accountability
3. Transformational capacity building
4. Transition focused leadership
5. Interlinked policy networks
6. Knowledge sharing communities.

While there is no international consensus regarding a specific form or universalised composition of a good society, there are, nevertheless, established shared visions of good states. For instance, good states maintain and respect borders, preserve sovereign authority, and protect their territories and peoples from threat (Edmondson & Levy, 2008). Good states contribute to international order, utilise their capacities to exercise authority over territory and citizens, and participate in international political decision-making (Edmondson & Levy, 2008; Krasner, 1999). In the twenty-first century, these principles are extending to normative expectations that good states balance their assessments of potential gains and risks in ways that take account of possible risks to others and other species.

The international political community is at a new juncture in relation to the moral capacities of states. Aspirations of sustainable development and adaptation to climate change rely upon ideas of morality and associated distributions of rights and responsibilities. Without shared moral visions supported by international law, the international political community will be unable to achieve just distributions of resources or maintain just institutions (Buchanan, 2007; Posner, 2003; Roberts & Parks, 2007). In the twenty-first century, as states grapple with intensifying issues arising from the impacts of global climate change and mass population movements, their moral capacities are simultaneously

challenged and expressed in new ways. Advocates of international justice find encouragement in observing how climate change impacts have
presented new ethical debates that have led to matters of intergenerational protection and management of territory and resources becoming
common themes in international political debates (Lind, 1995; Low &
Gleeson, 2001; O'Neill, 2009).

Ross Mittiga's examination of consumption-based carbon accounting
and the polluter-pays principle for the allocation of climate action burdens establishes that multidisciplinary knowledge can reduce uncertainties in responding to global climate change and that accountability is a
key element of fairness. Current and ongoing considerations about how
to allocate the costs of responsible climate action has been typified by
the use of fairness arguments by states to avoid bearing costs or to shift
costs to others. Allocating these burdens using economic mechanisms,
such as the polluter-pays principle and consumer-based accounting
approaches, can increase accountabilities but each approach has different impacts, especially in terms of achieving good environmental outcomes. Consumer-based accounting approaches provide an alternative
that can break away from some of the problematic burdens distributions
of polluter-pays principles—primarily because consumption is more
closely aligned with affluence.

There are no perfect economic approaches or mechanisms for
achieving improved accountabilities, better environmental outcomes,
and fairer distributions of burdens among states. Both polluter-pays
principles and consumer-based accounting are second best methods.
Nonetheless, consumer-based accounting has advantages in fairness,
environmental efficacy and cost-effectiveness, and consequently, these
attributes afford advantages in accountable climate governance because
states have fewer reasons to avoid full disclosure. Importantly, Mittiga
reminds us that climate mitigation is a process, rather than an outcome,
and that good climate governance mechanisms need to be future-oriented. Present- or past-oriented burdens allocating approaches contribute to future unfairness and also limit environmental benefits. Climate
mitigation involves an extended phase towards sustainable societies

that can be progressed through the contributions of ethics, law and economics.

Martina Grecequet, Jessica J. Hellman, Jack DeWaard and Yudi Li, in their comparison of human and non-human migration governance under climate change, determine that learning from observations of non-human species migrations can build capacities to move forward from current human migration governance approaches which were developed for very different reasons in the twentieth century. Climate change induced environmental transformations cannot be governed, and migration is an adaptation strategy that is currently underway by humans and non-humans alike. These migrations by human and non-human, and redistributions of non-human species, will both directly and indirectly affect where and how humans can live.

Barriers to migration increase the risks of mass extinctions of both human and non-human and reduce the effectiveness of migration as an adaptation strategy. Transdisciplinary knowledge can help to reduce these risks and identify and predict which humans and non-humans are vulnerable to particular types of migration barriers. Sustainability will only be achieved with better understandings of the nexus between human–non-human migrations, the nature–society nexus, and with specific attention to trapped populations in migration policies. To become more effective, climate governance mechanisms will need to include binding and non-binding instruments in updating migration policies to recognise that migration is an adaptation response to climate change induced environmental transformation.

Timothy Cadman, Tek Maraseni, Hugh Breakey and Hwan-ok Ma examine stakeholder perceptions and interests in climate governance through a detailed and systematic study of participants perceptions of their roles in key climate governance discussions. Their work sheds new light on understanding how accountability might be promoted in climate governance mechanisms. They determine that good accountable governance relies upon inclusiveness that provides stakeholders with access to influence and weight in evaluating decision- and policy-making. They point out that accountable, legitimate and good governance relies upon inclusiveness supported by resources, access to influence and weight in decision settings.

Their evidence suggests the international policy-making community is learning from experiences of climate change as multi-scale social and environmental issues. For instance, the Sustainable Development Mechanism of the Paris Agreement 2015 provides new attention to the accountability-mitigation nexus and new expectations are emerging that market linkages should also contribute to effective climate governance. These findings suggest that the realities of enmeshed globalisation and investment-driven trade relationships can become positive influences for climate governance in developing policies and accountability mechanisms. Finally, Cadman, Maraseni, Breakey and Ma show that established measures of good governance can be utilised to establish criteria for evaluating, and perhaps also predicting, the effectiveness of climate governance mechanisms. They conclude that inclusiveness is not sufficient to ensure all interests are represented, but accountability approaches can extend inclusiveness in climate governance.

Jeremy Moss examines the moral case for limiting fossil fuel exports and how the costs of climate governance are becoming higher because of delays in achieving effective mitigation and adaptation policies that respond to the challenges of climate-driven environmental transformations. He proposes that, as a mitigation strategy, restricting the supply of fossil fuels is an effective way to reduce contributions to climate change because it directly reduces levels of harmful greenhouse gas emissions. Regulating exports, which is an established mechanism in the international political economy, could be applied to fossil fuels as a morally defensible, simple and future-oriented strategy which could sit alongside other climate governance mechanisms without any inherent need for these to directly intersect.

He builds the case for this approach by acknowledging that the avoidance of harm is a central moral principle of many established governance mechanisms and that many governance mechanisms and institutions have a central purpose of ensuring morally grounded or morally defensible outcomes. States also are morally culpable entities that hold rights and responsibilities and ascribe these to others, which render them accountable for the harms they do and especially for the harms they knowingly do. Fossil fuel exporters are contributing to the risks of harm through supply, support and influence in affecting emissions outcomes, and knowingly exporting harm has international legal precedents that could be utilised to regulate in support of export restrictions. Supply-side approaches and constraints have advantages for climate governance in terms of more effective economic targeting and moral culpabilities. Moss argues that they could be speedily implemented by being incorporated into existing mechanisms and do not require new international governance architecture. A further advantage of this approach is that export restrictions can fairly distribute responsibilities for costs across producers and consumers.

Steve Vanderheiden examines the potential effectiveness of personal carbon trading and individual mitigation accountability as components of international climate change mitigation mechanisms. These would be scalable, hold agents responsible for their contributions to climate harms, could include equity and ambition goals, and be applied to both individuals and states. These approaches would be particularly valuable given that the Nationally Determined Contributions required under the Paris Agreement are not legally binding but rely upon reputational accountability and will not currently meet 'safe' climate change targets of limiting global warming to 2°C or lower.

Making individuals more accountable for the greenhouse gases their conduct produces will also make states more accountable and doing so builds upon existing accountabilities between states and citizens. The use of personal carbon budgeting could extend the responsibilities factored into personal carbon taxes, which are already being adopted in some states, and more effectively promote just distributions of costs, reduce emissions, and fund further decarbonising initiatives including new energy opportunities. By promoting cognitive responsibility, which is a key component in accountabilities, personal carbon budgeting establishes a central principle of collective sacrifice and thereby provides a short cut through complex negotiations regarding the allocation of burdens and preferred models of fairness. The promotion of governmental accountability and personal responsibility then provides a powerful political and economic nexus that can improve environmental outcomes and promote sustainable societies.

References

Buchanan, A. (2007). *Justice, legitimacy, and self-determination: Moral foundations for international law*. Oxford: Oxford University Press.

Edmondson, B., & Levy, S. (2008). *International relations: Nurturing reality*. Frenchs Forest: Pearson Education.

Krasner, S. (1999). *Sovereignty: Organized hypocrisy*. Princeton: Princeton University Press.

Lind, R. C. (1995). Intergenerational equity, discounting and the role of cost-benefit analysis in evaluating climate policy. *Energy Policy, 23*(4), 379–389.

Low, N., & Gleeson, B. (2001). The challenge of ethical environmental governance. In B. Gleeson & N. Low (Eds.), *Governing for the environment: Global problems, ethics and democracy*. Houndmills: Palgrave.

O'Neill, K. (2009). *The environment and international relations*. Port Melbourne: Cambridge University Press.

Posner, R. (2003). *Law, pragmatism and democracy*. Cambridge, MA: Harvard University Press.

Roberts, J. T., & Parks, B. C. (2007). *A climate of justice: Global inequality, north-south politics and climate policy*. Cambridge: MIT Press.

8

Allocating the Burdens of Climate Action: Consumption-Based Carbon Accounting and the Polluter-Pays Principle

Ross Mittiga

1   Introduction

That human activity is causing profound and potentially catastrophic climate change is no longer a matter of serious debate among climate scientists. By century end, average sea levels may rise as much as 2.5 m (8 ft.), displacing millions living on the coasts.[1] Warming will exacerbate droughts, flooding, heat waves, and soil aridification, all of which seriously threaten agriculture. Zika, dengue, malaria, cholera, and other mosquito-borne illnesses will proliferate as hotter climates expand the insect's range. Warmer winds and water portend more powerful and frequent storms, and thus increased strains on critical infrastructure.

---

[1]National Oceanic and Atmospheric Administration et al., 2017. Cf. IPCC, AR5, WG2, pp. 366, 368–369, which predicts 0.98 m.

R. Mittiga (✉)
Instituto de Ciencia Política, Pontificia Universidad Católica de Chile, Santiago, Chile
e-mail: ross.mittiga@uc.cl

© The Author(s) 2019
B. Edmondson and S. Levy (eds.), *Transformative Climates and Accountable Governance*, Palgrave Studies in Environmental Transformation, Transition and Accountability, https://doi.org/10.1007/978-3-319-97400-2_8

Developing fair, effective, and accountable responses to these threats is essential. Central here is the question of how to allocate the costs of climate action among states (Page, 2011, p. 413).[2] Indeed, for the last thirty years, this has been *the* question of climate politics at the international level. While virtually everyone agrees that the distribution of costs should be fair, there remains serious disagreement about what constitutes fairness. This disagreement follows from a vague but critically important provision in the U.N. Framework Convention on Climate Change (or UNFCCC), which holds that states 'should protect the climate system ... on the basis of equity and in accordance with their common but differentiated responsibilities and respective capabilities' (UNFCCC, 1992). Political leaders have exploited the ambiguity here, invoking fairness every time they find a policy or treaty too stringent for themselves, or not stringent enough for others. For example, after being criticized for its less-than-ambitious emissions targets, India—the world's third largest emitter of greenhouse gases (GHGs)—argued that (1) at least on certain metrics, Indians have contributed very little to the climate problem and (2) even if they are responsible in a collective sense, expecting further action would be unfair, insofar as it would impede their ability to achieve critical development gains.[3] Of course, India is not the only country to advance claims like these: Many poor and developing states invoke a 'right to grow' or develop when pressed for greater climate action.

Wealthy, developed states likewise appeal to fairness to justify climate inaction—though their claims are clearly more duplicitous. The U.S. Senate, for instance, cited fairness as one of its reasons for refusing to ratify the Kyoto Protocol (Senate, 1997). More recently, President

---

[2]Throughout, I assume that states are the relevant duty bearers. For a critical discussion of who should bear climate duties, see Caney (2005).

[3]Quoting directly: 'Both in terms of cumulative global emissions and per capita emissions, India's contribution to the problem of climate change is limited but its actions are fair and ambitious. ... Nations that are now striving to fulfill the 'right to grow' of their teeming millions cannot be made to feel guilty [about] their development agenda' (India's Intended Nationally Determined Contribution, pp. 1, 33).

Donald Trump invoked fairness concerns to defend withdrawing from the Paris Accord (Trump, 2017).[4]

The debate among states closely tracks scholarly debates over fairness in the context of climate change. Although this debate is still ongoing, climate ethicists have largely converged upon three main principles:

- The polluter-pays principle (PPP): Those responsible for causing climate change should pay, in proportion to their contribution.
- The ability-to-pay principle (APP): The wealthy should pay, in proportion to their wealth.
- The beneficiary-pays principle (BPP): Those who have benefitted from activities that cause(d) climate change should pay, to the extent they have benefitted.

Of these, the PPP is widely regarded as the most intuitively plausible and well established in international environmental law (see Sect. 2). In recent years, however, scholars have subjected the PPP to extensive criticism, for reasons examined in Sect. 3.

One avenue to resuscitate the PPP entails taking into account consumption emissions—i.e., embedded in global trade flows. Although others have proposed adopting consumption-based emissions accounting, there has been no attempt to connect this change to the PPP—or any other distributive principle. Moreover, virtually no attention has been paid to the ethical justifications for holding consumers, rather than producers, responsible—a point addressed in Sect. 4. By adopting a consumption-based emissions accounting method, the resultant distribution of burdens closely tracks economic capacity without resorting to problematic attributions of historical responsibility, as standard formulations of the PPP do. This change also offers a way to address emerging problems such as carbon leakage.

---

[4]Specifically, Trump (2017) said: 'The bottom line is that the Paris Accord is very unfair, at the highest level, to the United States. ...I will work to ensure that America remains the world's leader on environmental issues, but under a framework that is fair and where the burdens and responsibilities are equally shared among the many nations all around the world.'

Before proceeding, it is helpful to explain what the 'climate burdens' are that must be allocated. The literature typically elucidates two, though a third is increasingly recognized.[5] First is the duty of mitigation, which involves reducing greenhouse gas emissions *and* enhancing natural 'sinks,' which absorb and store or convert emissions into non-insulating chemicals. We fulfill this duty by reducing energy usage or adopting carbon-free forms of energy production, supporting the development of green infrastructure (e.g., through technology transfers), consuming fewer animal products (Wellesley, Happer, & Froggatt, 2015, p. vii and passim; Steinfeld et al., 2006), travelling less in motorized vehicles (especially aircraft), preventing deforestation, and promoting afforestation.

A second duty is that of adaptation, which involves helping people (and perhaps other animals) adjust to climate changes.[6] We fulfill this duty by promoting access to vaccinations, constructing seawalls, and developing infrastructure (like water pumps and levees) to manage flooding and drought (Eckersley, 2015).

Finally, there is the duty to provide compensation for the 'adverse effects of climate change that cannot, or will not, be prevented through policies of mitigation or adaption' (Page, 2016, p. 84).[7] What this duty entails in practical terms remains controversial. At minimum, though, it requires the establishment of an international mechanism—like a risk-pooling insurance scheme (Arrow, Parikh, & Pillet, 1995, p. 72) or direct-aid fund—capable of providing support to states or people affected by climate change. The Warsaw Mechanism is a first step in this direction (James et al., 2014, p. 938).

Most agree that all three duties are essential. Mitigation is necessary, for instance, to prevent the crossing of 'nonlinear threshold points' (Caney, 2010, p. 205; Gardiner, 2004, p. 562)—also known as

---

[5]In particular, since the 18th Conference of the Parties (COP 18), in 2012.

[6]The IPCC defines adaptation as any '[a]djustment in natural or human systems in response to actual or expected climatic stimuli or their effects, which moderates harm or exploits beneficial opportunities' (IPCC, AR4, WG3, 18.1.2). See Jamieson (2010, pp. 265–266).

[7]Specifically, policy-makers define loss as 'negative impacts of climate change that are permanent' and damage 'as those impacts that can be reversed' (Huq, Roberts, & Fenton, 2013).

'tipping points'—'beyond which major changes occur that may be self-reinforcing and are likely to be irreversible over relevant time scales' (Furman et al., 2014, p. 20). Tipping points are unpredictable and very dangerous. Some tipping points, like the release of methane in the northern hemisphere's (already thawing) permafrost, risk positive feedback cycles that could generate 'runaway' global warming.[8] Given this, reducing emissions and enhancing sinks are essential; however, climate change has already progressed passed the point that all harmful changes can be avoided by mitigation alone. Because GHGs remain in the atmosphere long after they are released, even sharp emissions cuts now will not prevent global temperatures from continuing to rise well into the future (Caney, 2010, pp. 204–205). Adaptation is thus also necessary to avoid grave threats to plant, animal, and human life. Yet, there are many climate changes and events that will exceed our anticipatory adaptation capacities; thus, establishing a compensation fund for losses and damages is also crucial.

Fulfilling these three duties is quite costly. On some estimates, for instance, effective mitigation alone would cost around $780 billion (in 2015 USD$) every year, for the foreseeable future (Stern, 2007, pp. 258–262).[9] Hence our original question: according to which principle(s) should we allocate climate-action burdens? A satisfactory answer must be *comprehensive*—able to cover all three action burdens effectively, now and into the future—and *fair*—sensitive to differences in states' contributions to the problem and their differing capacities to deal with it. Anything less will fall short of the demands of accountable and effective climate governance.

In their standard formulations, none of the three principles cited above satisfy both of these desiderata. Yet, by reformulating the PPP to take account of emissions embodied in global trade flows, we can get close. Remaining shortcomings, outlined in Sect. 5, can be overcome by

---

[8]One analysis suggests that a large-scale methane release could generate $60 trillion in damages (Wagner & Weitzman, 2016, p. 185).

[9]For similar estimates, see Stern (2010, p. 45), Weitzman (2007, p. 720), Nordhaus (2009, p. 90). For discussion, see Caney (2009, p. 182, *n.* 9), Page (2011, p. 412), and Rendall (2011, p. 890).

supplementing the principle to produce a pluralist, bi-phasic theory of distributive climate justice that is fairer and more environmentally effective than alternatives.

2   The Polluter-Pays Principle

Many believe that those who cause harm or damage should (pay to) fix it (or compensate for any resultant suffering). Perhaps for this reason, the PPP is considered highly intuitive and has been a fixture of international environmental law well before climate change was recognized as a major problem.[10]

The principle first appears in the climate-change context in Principle 16 of the Rio Declaration on Environment and Development, which states: 'National authorities should endeavor to promote the internalization of environmental costs and the use of economic instruments, taking into account the approach that the polluter should, in principle, bear the cost of pollution.' Notably, this formulation is (1) present oriented[11] and (2) focused primarily on reforming the behavior of economic actors. It calls on governments to require agents under their jurisdiction to include any negative environmental externalities in the price of their goods. This economistic formulation of the PPP is used widely. Nicholas Stern, for instance, argues that continued growth of greenhouse gas-emitting activities represents a major market failure, the main solution for which is to force agents to shoulder the social costs of their emissions by '[p]utting an appropriate price on carbon' (Stern, 2007, p. xviii). Notably, this way of formulating the PPP is present oriented, which simply means that it is focused on *current* pollution only—on taxing the GHGs being emitted in here and now (or at some regular interval, such as annually).

---

[10]An early example is the OECD's 1972 *Recommendation of the Council on Guiding Principles* (OECD, 12). See also, Article 130R of the Maastricht Treaty, the Commission on Global Governance, and IPCC, AR5, WG3, 217–218, 318, 1268.

[11]While the literature commonly refers to principles focusing on current emissions as 'forward-looking' (Shue, 1999, p. 534), to allay confusion, I opt instead for the term 'present-oriented.'

The present-oriented PPP is not only economically valuable, as a mechanism for eliminating inefficiencies (Broome, 2012); as a principle of liability, it also captures an important part of moral and political duty. Morally speaking, we want agents to get what they deserve. If someone harms or endangers others, we typically believe that agent should be held accountable—and this falls to government. As John Rawls explains, making agents pay for the 'full social cost of their action' is an 'essential task of law and government' in a just society. Hence, he defends a present-oriented PPP, pointing to 'striking cases of public harms,' as when industries sully and erode the natural environment.' In such cases, Rawls argues, government must correct the 'divergence between private and social accounting that the market fails to register' (Rawls, 1999, p. 237). In practice, this means pricing emissions in a way that reflects the harm they are causing *and* some estimate of the harm they may cause (in the form of a risk premium).[12] Such a tax would provide a 'double dividend'[13]—it would curb emissions (by making them more expensive) while providing a revenue stream for financing adaptation and compensation efforts—making the principle an attractive basis for international climate policy.

3   The Case for, and Problems with, a Backward-Looking PPP

Despite its appeal, the present-oriented PPP has attracted trenchant criticism in recent years. Henry Shue, Eric Neumayer, Simon Caney, and others object to its neglect of historical emissions, arguing that agents most responsible for the emergence of climate change owe a debt of corrective justice to those adversely affected by it (Shue, 1999, p. 534ff). These scholars also contend that a purely present-oriented PPP unduly burdens developing states, which rely on emissions-heavy industrialization to sustain minimal standard-of-living. For these

---

[12]I elaborate this point below.
[13]See Caney (2010, n. 31; but cf. Stern, 2010, p. 62).

reasons, they endorse backward-looking PPPs (Caney, 2010; Neumayer, 2000, pp. 185–192; Moellendorf, 2012),[14] which allocate climate duties in proportion to cumulative (historical) emissions.

Concerns that a purely present-oriented PPP would heavily burden developing states are not unfounded.[15] Using standard accounting methods, six of the top ten emitters in 2012 were developing states,[16] and nearly two-thirds of all emissions came from developing and poor states. Thus, insofar as a present-oriented PPP ignores historical emissions, it manifests a compound unfairness: It forgives post-industrial states their harmful historical emissions while placing heavy burdens on those least able to bear them. On this view, corrective justice and distributive justice are better served by including historical emissions in assessments of responsibility.

3.1   Disappearing Perpetrators

As others have noted, however, this solution is practically and theoretically fraught.[17] We can note two commonly cited issues here. First, many historical polluters are now dead and therefore cannot be made to pay. Forcing their descendants to pay, as Edward Page notes, violates the 'ethos' of the PPP, which 'presupposes that only agents that actually caused an environmentally adverse outcome can be held' responsible (Page, 2011, p. 415). This is the 'disappearing perpetrators' problem. A commonly proposed solution to this problem is to hold temporally unbounded actors like states responsible. Doing so raises new issues. First, many states have undergone one or more revolutions since

---

[14]Note that Shue (1999) does not refer to his principle as a PPP, but—as Caney (2005, p. 753) notes—it fits the mold.

[15]The following is based on measures using the CAIT data set.

[16]Specifically, China (1), India (3), Russia (4), Indonesia (5), Brazil (6), and Mexico (10). This ranking includes emissions from land-use changes and counts any state with a per capita GDP below USD $12,000 as 'developing.' Notably, several developing states also top the list for per capita emissions and post-1990 emissions growth.

[17]For further critique of backward-looking principles, see Kingston (2014). See also Shue (1999), Singer (2002), Caney (2005, 2010), and Page (2011, 2016).

industrializing. Should states with new constitutions or leadership be responsible for the actions of the regimes they supplanted?[18] Similarly, what of former colonies? (Should emissions generated in Ghana before 1957 be attributed to the UK or the current government of Ghana?) Moreover, unless we take the implausible view that states have agency distinct from the human authorities directing them, it also seems problematic that many former authorities are dead. (For instance, the most intense deforestation in the USA occurred between 1850 and 1920 (MacCleery, 2011, p. xii); are current citizens responsible for those land-use changes, despite having no part in authorizing them?) In short, making present generations responsible for historically remote emissions fails to satisfy the central dictum 'the polluters should pay' (Posner & Weisbach, 2010, pp. 108–109).[19]

3.2   Excusable Ignorance

Until the problem of climate change became firmly established scientifically and widely known, all agents—including states—could be said to have been acting in 'excusable ignorance' of the harm their actions (particularly emitting GHGs) were causing. In response to this, Peter Singer (2002, p. 34), Eric Neumayer (2000, pp. 181, 189), Henry Shue (1999, p. 536), and others have proposed establishing a 'cut-off date' for excusable ignorance: a date after which knowledge of climate change was readily available and thus agents could be held responsible.[20] Most often, the date proposed is 1990, which is the year the first IPCC report was released.

---

[18]Page (2011, p. 415) argues that it would be inappropriate to hold new governments responsible for the actions of former regimes. See also Kingston (2014, p. 284ff), Caney (2006, p. 469ff), and Miller (2009, p. 151ff).

[19]Also problematically, many (and perhaps most) climate-change *victims* do not yet exist, since grounding corrective-justice claims usually requires establishing an identity between victim and wrongdoer.

[20]Singer and Shue have proposed 1990, while Neumayer suggests the mid-1980s. For critical discussion, see Caney (2005, pp. 762, 769), Page (2011, p. 415; 2016, p. 93).

Although this seems a compelling (if only partial[21]) solution, there are two complications. First, the disappearing perpetrators problem still applies, if in attenuated form. That is, many 'knowing' polluters in the period since 1990 have also died. Moreover, many of the people alive today are children or were for much of the time since 1990. Assuming we cannot hold people accountable for what happens while they are (or were) children, the number of fully culpable adults (i.e., those who were adults in the year 1990 and are still alive now) is quite small relative to all those who are alive now or were at some point between 1990 and today. If our aim is to make the polluters pay, these considerations must be taken into account when assigning responsibility—viz., we must determine how much of the global stock of atmospheric GHGs is attributable to actors no longer alive or who are or were children in the period from 1990 to today. This is a daunting, if not impossible, task. We might avoid these issues by designating states as the relevant agents, but this would raise anew many of the above issues.

Even setting these complications aside, it quickly becomes clear that focusing on post-1990 cumulative emissions does little to resolve the initial concern with the present-oriented PPP—viz., that it entails economically regressive burdens. Indeed, in 2012, five of the top ten states for *post-1990* historical emissions were developing economies.[22]

3.3   Modifying the Principle

Many have proposed revisions to the backward-looking PPP to address these and other issues. For instance, some advocate a principle of strict liability, arguing that agents should be held responsible for emissions whether or not they understood the consequences of their actions, or even could have known (Gardiner, 2004; Neumayer, 2000; Shue, 1999, pp. 531–545). Others charge that strict liability is morally questionable

---

[21]Some resist this move because it curtails the PPP's reach and thus its ability to serve corrective justice (Caney, 2010, p. 209; Page, 2016, p. 93).

[22]Specifically, (2) China, (3) Russia, (4) India, (7) Brazil, and (10) Mexico. NB: This is the list for emissions *excluding* land-use changes. The list *including* land-use changes is even more regressive.

(Caney, 2010; Kingston, 2014, pp. 287–288), however, and likely unserviceable as a basis for international climate policy (Baer, 2010, p. 248; Bell, 2011; Moellendorf, 2012; Posner & Weisbach, 2010; Schüssler, 2011).

Another solution involves importing a notion of 'benefitting' into the PPP which entails arguing that those who have gained 'unjustly' from historical emissions ought to pay. Caney, for instance, modifies his PPP to hold that 'if people engage in activities which jeopardise other people's fundamental interests ... they should bear the costs of their actions even if they were excusably ignorant [provided] *they have benefited from those harmful activities*' (Caney, 2010, p. 210). Shue similarly argues that 'current generations in affluent states with high historical emissions are, and future generations probably will be, continuing beneficiaries of earlier industrial activity'—and thus should pay (Gosseries, 2004; Shue, 1999, p. 536; see also Neuamyer, 2000, p. 189; Page, 2016). Whether or not this offers a coherent way forward,[23] it means abandoning the PPP for a BPP, and with it, the central moral intuition that those responsible for causing a problem should pay to address it.[24] If we are to preserve this intuition, we must determine whether an alternative formulation of the PPP is available—specifically, one that tracks a plausible notion of contribution while remaining sensitive to different states' economic capacities.

4   A Revised, Present-Oriented PPP

A properly formulated, present-oriented PPP can accomplish this. Recall that the present-oriented PPP stipulates that those who contribute to climate change should pay for climate action, in proportion to their contribution. Thus, the more emissions an agent generates, the greater that agent's burdens should be. Above, we noted the concern

---

[23]I suggest that it does not in Sect. 5.4 (ii).

[24]Caney (2005, p. 757) recognizes this in an earlier article, but does not register the point against himself in the piece quoted above (Caney 2010). For criticisms of the BPP, see Kingston (2014, p. 288ff). For a defense, see Page (2016).

that this entails imposing heavy costs on developing countries (for instance, China, India, and Mexico), which generate significant yearly emissions.

This concern is valid *if* we adopt 'production-based' emissions accounting, which is virtually ubiquitous: It forms the basis of the UNFCCC and Kyoto Protocol calculations and is almost always used in public discussions of national emissions totals (Davis & Caldeira, 2010; Peters and Hertwich, 2008b; Steininger et al., 2014). Perhaps because of this, production-based accounting (PBA) has attracted little critical attention from climate ethicists.[25] It is not the only accounting method, nor the most normatively compelling.

An increasingly recognized alternative—consumption-based accounting (CBA)—traces emissions 'embodied'[26] in trade goods and attributes responsibility for those emissions to the country in which the goods are consumed. So, for example, using CBA, emissions generated in China to produce goods consumed in Norway are attributed to Norway. This small modification helps reveal often obscure neocolonial relations, whereby rich and powerful states outsource the production of goods to countries with cheaper labor markets, and then blame those countries for having higher emissions profiles. Chinese officials and environmental advocates have expressed particular frustration with this. For instance, at a press conference, Qin Gang, China's Foreign Ministry spokesman, once reminded Western news outlets that a 'lot of what you use, wear and eat is produced in China…On the one hand, you increase production in China; on the other hand you criticise China on the emission reduction issue' (Scientific American, 2018). Similarly, Yang Ailun of Greenpeace China claims that, in the last 30 years, '[a]ll the West has done is export a great slice of its carbon footprint to China and make China the world's factory' (Scientific American, 2018).

Taking these outsourced emissions into account could radically transform how we understand state responsibility. As Davis and Caldeira

---

[25]Some in the climate-policy community have raised fairness concerns, however. See Steininger et al. (2014, 2016), Davis and Caldeira (2010), Kander et al. (2015).

[26]IPCC, AR5, WG3, 306; Davis and Caldeira (2010).

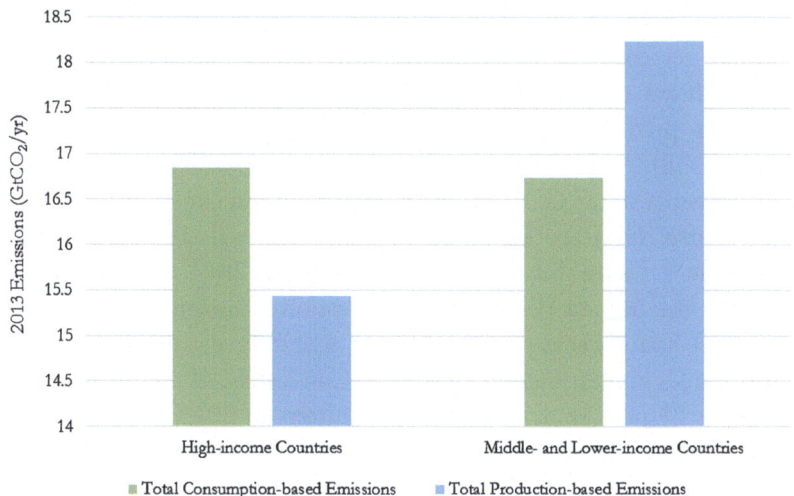

Fig. 1 Distribution of consumption—and production-based emissions

note: 'Approximately 6.2 gigatonnes (Gt) of CO_2, [or] 23% of all CO_2 emissions from fossil-fuel burning [in 2004]…, were emitted during the production of goods that were ultimately consumed in a different country' (Davis & Caldeira, 2010, p. 5688). Consequently, CBA provides a very different picture of national emissions than PBA.

We can see this in the following graph. In Fig. 1, the difference between the blue and the green columns for high-income countries, on the one hand, and middle- to low-income countries, on the other hand, shows that many of the emissions produced by the latter are embodied in goods consumed by the former.[27] This 'off-shoring' of emissions— from rich to poor and developing countries—is made possible through international trade.

Given that the biggest importers of goods from the developing world are affluent states, we should expect them to top the list for

---

[27]This follows the World Bank's state income grouping scheme, in which countries with a GNI per capita above USD$12,475 are considered high income. https://datahelpdesk.worldbank.org/knowledgebase/articles/906519.

consumption-based emissions. The data show precisely this.[28] In fact, in a data set covering 119 countries[29] for the year 2013,[30] regression analysis shows that a state's GDP was a very strong and statistically significant predictor of its total consumption emissions ($R^2 = 0.76$; $p = 3.62E\text{-}38$).[31] Notably, this relationship is appreciably stronger than that between a state's GDP and its total production emissions.[32]

Of course, GNI (or GDP) per capita better reflects affluence than GNI alone.[33] Using this metric, we find that, in 2013, of the 25 countries with the highest per capita consumption emissions, all (save one, Mongolia) had a per capita GNI above \$12,475, placing them in the World Bank's highest income bracket.[34] Conversely, of the states among the lowest 25 per capita consumption emitters in 2013, *none* had a per capita GNI exceeding \$2,700 (the average was \$1,030.40).[35] In short, the countries with the highest per capita consumption emissions are also the wealthiest.

Furthermore, the top 25 per capita consumption emitters in 2013 were responsible for 33.46% of all emissions generated that year,[36] despite representing only 10.64% of the global population.[37]

---

[28]Emissions data from Peters et al. (2011). Economics and demographic data from World Development Indicators (last updated March 1, 2017), supplemented, to include Taiwan, by https://eng.stat.gov.tw/ct.asp?xItem=37408&CtNode=5347&mp=5. All economic information is reported in 2016 USD\$, unless otherwise stated.

[29]NB: Poor countries are under-represented in this data set—a tendency among most CBA data sets. If they were fully included, the results would likely be even more striking.

[30]At the time of writing, 2013 is the most recent year for which most data are available.

[31]Similar results are found for GNI; $R^2 = 0.74$, $p = 1.32E\text{-}35$.

[32]The data do report a statistically significant correlation between GDP and production-based emissions ($p = 8.42E\text{-}30$), but with less explanatory power ($R^2 = 0.67$).

[33]China and India, for instance, have high GDPs/GNIs but are still relatively poor because of their large populations. Because of this, the World Bank uses a per capita measure as the basis of its income groupings.

[34]The average per capita GNI of this group was \$42,863.44—almost triple the world average at the time, \$14,928.37.

[35]Again, this is even more striking given that the data set excludes many poor and developing states.

[36]The consumption emissions of these states totaled 12,040.71 $MtCO_2$, while global emissions in 2013 totaled approximately 35,986.20 $MtCO_2$.

[37]Or 760,609,121 of 7,147,749,368 people.

Comparatively, the bottom 25 per capita consumption emitters, which represent 14.40% of the global population,[38] were responsible for just 1.62% of all emissions.[39] This, again, strongly suggests that consumption emissions closely track affluence.

Of course, using a consumption-based emissions accounting method effectively means shifting responsibility from producer to consumer because the underlying premise of the CBA model is that consumer demand (particularly in affluent states) is the proper locus of responsibility. There are several reasons for accepting this view.

For one, it is for the sake of satisfying consumer demand that producers undertake emissions-generating activities. Now we might worry that while consumer demand gives the reason for production, it does not, all else being equal, say anything about how production should be conducted—especially, whether producers should employ higher- or lower-emission productive processes. That decision, it seems, falls to producers. Yet, this might be too hasty. A basic axiom of supply and demand (and rational decision theory) is that when deciding between identical goods, consumers will, all else being equal, prefer whichever is cheapest. There are of course exceptions, carved out by marketing, ethical considerations, and so on. However, in most cases, price is the surest determinant of demand. If this is right, then consumers are, in effect, driving a specific kind of production: namely, cheap—which often means emissions intensive—production. In other words, consumer preferences for cheap goods dictate—in a general, but direct and significant way—producers' decisions about *how* they produce goods.

There are two additional reasons to hold consumers responsible. Affluent, consumerist states are the chief architects of today's highly liberalized trade system and global division of labor, which enables multinational firms to offshore high-emitting productive processes. In other words, the reason why production proceeds in emissions-insensitive ways is, in an important sense, because the political leadership of

---

[38]Approximately 1,028,951,328 people.
[39]That is, about 581.86 $MtCO_2$.

consumerists societies has structured it that way, to ensure an unimpeded stream of cheap goods. Finally, locating responsibility with consumers gives practical expression to the widely shared idea that successfully combatting climate change requires the wealthy to undertake meaningful lifestyle changes.[40] Holding consumers responsible will have the result of curbing consumption.

These points notwithstanding, clearly producers bear *some* responsibility. For in many cases, they have the ability, if not the will, to lower their emissions by adopting alternative modes of production. Yet, even granting this, two considerations should lead us to be wary about treating producers as equally responsible as consumers. First, in many if not most cases, producers lack an incentive to reduce their emissions unilaterally, for doing so would raise the price of their goods and thus decrease their market competitiveness. Of course, governmental regulation offers a potential solution to this dilemma: By requiring all productive firms in a given territory to lower their emissions, none will be made less competitive on that ground. Yet, as Fig. 1 shows, we may worry that at precisely the moment stringent emissions-reduction policies are implemented by any given state, multinational firms will simply relocate to a country without such restrictions. This dynamic is a perverse feature of our global trade system, which, again, was designed largely *by* and *for* affluent states to ensure access to cheap consumer products.

Still, conscientious producers could, for example, use advertising to instill consumer demand for lower-emissions goods (much as organic agricultural suppliers have done for pesticide-free produce), or undertake lobbying efforts to create industry-wide regulations, eliminating perverse incentives from the decision-making process. To the extent, then, that producers have failed to inform consumer preferences or reform the system, they appear to be fit objects of blame. A problem with this view, though, is that successful productive firms are often directed by agents who, themselves, are high consumers—for

---

[40]On this, see IPCC, AR5, WG3, 6.8, 7.9, 8.3.5, 8.9, 9.2–9.3, 9.10, 10.4, 11.4, 12.4–12.7, 15.3–15.5.

consumption largely tracks wealth at the individual level, just as it does on the international level.[41] This suggests another layer of perverse incentives. Calling on the leaders of productive firms to reform themselves or their industries is, in effect, calling on the most prodigious global consumers to make their own consumption more expensive.

Taking all of this into account, it seems that, on balance, there is a stronger case for holding consumers responsible. Doubtlessly, though, there are relevant counter-examples and exceptions in particular cases and contexts. For this reason, we might conclude that the best accounting method would be one capable of isolating the relative causal impact of consumers and producers, and assigning 'shares' of responsibility. Parsing the causal impacts of consumers versus producers, however, is—philosophically and practically—infeasible (Steininger et al., 2014, p. 78).

We must therefore pick between second-best methods and attribute responsibility accordingly. In other words, choosing between PBA and CBA is less a matter of which method better captures contribution (which both do imperfectly), but of which performs better with respect to pertinent ethical factors: especially, fairness, environmental efficacy, and cost-effectiveness. I argue that CBA is superior in these three regards and thus provides the most ethically compelling basis for allocating climate burdens.

4.1 Fairness

As already noted, CBA and PBA offer starkly different views of national emissions: CBA provides a lower-emissions count for lower- and middle-income countries, and a proportionally higher one for affluent states.[42] In this way, CBA satisfies what we might call the convergence view in the literature: that the rich ought to pay for climate action (Gardiner, 2004, p. 579). For, as shown above, a significant and robust correlation exists between GNI (per capita) and consumption emissions

---

[41]It is for this reason that Rawls (1999, pp. 199, 246) prefers consumption-based over income-based taxation at the domestic level.

[42]This finding is affirmed also in the most recent IPCC report (AR5, WG3, 127).

(per capita). This relationship is stronger on a present-oriented CBA model than a PBA model. Moreover, the same data used above show that there is no statistically significant relationship between per capita GNI and (post-1990) historical production-based emissions.[43] In other words, contemporary affluence is *not* a good predictor of cumulative production emissions. This suggests that those who propose counting historical emissions to satisfy the convergence view would do better to advocate a present-oriented, consumption-based PPP.

One might object that imposing a border tax (or tariff) in accordance with a present-oriented CPP would unfairly harm those in developing states, which are net exporters. I address several versions of the unfairness objection in Sects. 5.1 and 5.2. We can address the trade objection here, however. This objection can be more formally stated as follows:

a. *If* universally (or nearly universally) implemented greenhouse gas taxes are necessary, as most believe, to combat climate change successfully[44];
b. and *if* such taxes necessarily dampen trade, thereby negatively affecting developing states,
c. *then* we must choose between temporarily depressing international trade and addressing climate change.

There are strong reasons to doubt (b), or at least the provision that depressed trade must come at the cost of those in developing states. Through international aid programs, guaranteed minimum trade deals, international investments in green infrastructure, etc., any negative trade effects brought on by greenhouse gas taxes could be significantly mitigated or even reversed.

Yet, even granting (b), we should note that failing to address climate change now will almost certainly depress economic activity in the future, especially in poor and developing states, which are

---

[43]Regression analysis between GNI per capita and cumulative production emissions since 1990 returns a *p*-value of 0.18—far below any significant threshold.

[44]I defend this point in the next subsection.

disproportionately vulnerable to environmental changes. Furthermore, this harm would likely far exceed any foregone welfare gains related to greenhouse taxes.[45]

Suppose then that we are justified in imposing greenhouse gas taxes today.[46] The question then becomes: Is it fairer to place greenhouse gas taxes directly on producers or consumers? Although under ideal conditions there should be no difference, in our fragmented global system, placing taxes on consumption will more likely ensure that the rich pay—because, again, net consumption (consumption in excess of production) increases in rough proportion to wealth. Thus, a consumption-based model appears to be economically fairer.

4.2   Environmental Efficacy

Understanding environmental efficacy in terms of a principle's or policy's ability to 'reduce the causes and impacts of climate change,'[47] CBA has a crucial advantage: It can help prevent 'carbon leakage.' Carbon leakage, in both 'strong' and 'weak' variants, is the by-product of international free-trade agreements and fragmented mitigation policies. Strong leakage occurs when unilateral mitigation policies prompt domestic, polluting firms to relocate offshore.[48] Weak or 'consumption-induced' leakage arises from the global division of labor, which, today, concentrates production in states with energy-inefficient infrastructure.[49] In both cases, reductions in emissions in one state are offset by equal or greater increases in another.

Carbon leakage poses a problem for PBA so long as mitigation policies are not unified and there is not a single price for carbon enforced

---

[45]I return to this point below, via Rendall (2011).

[46]I provide a much fuller defense of this point in Mittiga (2018).

[47]IPCC, AR5, WG3, pp. 1009, 236.

[48]According to the IPCC: 'Carbon leakage is...the increase in CO_2 emissions outside the countries taking domestic mitigation action divided by the reduction in the emissions of these countries' (AR4, WG3, Chapter 11.7.2).

[49]On weak vs. strong, see Steininger et al. (2014, pp. 76, 79ff), Peters and Hertwich (2008a), Droege (2011), IPCC, AR5, WG3, p. 386.

internationally. For whenever emissions-producing activities are relocated to states with less stringent policies, regulation is undermined. To clarify, imagine state X imposes and strictly enforces a tax on greenhouse gas emissions, while state Y does not. Assuming a lack of trade barriers between X and Y, we can expect that X's policy will induce some heavily polluting companies to relocate to Y, and simply export back the goods they produce there. In this way, the emissions X sought to curtail continue, unabated. Certainly, PBA can (and will) reflect lower national emissions for X, but the aim is to reduce *global* emissions, not those of any particular state (except in an intermediate sense).

Even within a fragmented system, CBA can combat this: Through the medium of international trade, state X can enforce mitigation policies (such as a price on GHGs) outside its borders by applying tariffs to imported products. Economists call such tariffs 'border carbon adjustments'; they work by adjusting imported-product prices to reflect the social costs of greenhouse gas emissions embodied within them (Brooks, 2015; Steininger et al., 2014, p. 76). By subjecting all goods exported from states that lack (sufficiently stringent) mitigation policies to emissions taxes as a condition of market participation, CBA is able to deter free-riding. By minimizing or preventing carbon leakage in this way, CBA increases the efficacy of mitigation efforts undertaken by any trading state.

4.3 Cost-Effectiveness

By 'bringing the export sectors of the developing and emerging economies' that affluent states trade with 'into the scope of [the latter's] policy,' CBA also has an advantage in cost-effectiveness (Steininger et al., 2014, p. 81). In other words, CBA is able to capture a broader share of global emissions and because the costs of mitigation increase with the fraction of total emissions abated, this means that CBA will help identify cheaper mitigation targets. This is an example of declining marginal costs: It is cheaper and easier to reduce emissions in countries that have not decarbonized much or at all—which is the case in many developing states—and more difficult and expensive to reduce emissions in states that are already decarbonizing—which is the case for many affluent states (Barrett, 1998; Steininger et al., 2014, p. 81; Stern, 2007). In other words, by making consumers responsible for the emissions embodied in trade goods, we incentivize emissions-reduction efforts in net-exporting countries, which can often be less costly than comparable reductions efforts in net-importing countries.

4.4 Additional Considerations

A present-oriented, consumption-based PPP offers two further advantages. First, it is sensitive to changing economic fates. As developing states become wealthier, it is essential that they commit more resources to combatting climate change. The consumption-based PPP can explain why, and to what extent, they should contribute. A backward-looking principle cannot do this; nor can it accommodate economic decline. For on a backward-looking principle, present circumstances matter little if at all. Insofar as consumption emissions decrease in accordance with economic capacity, a consumption-based, present-oriented PPP offers security: Should the rich become poor, or the poor rich, burdens change accordingly.

Second, as Davis and Caldeira note, 'to the extent that constraints on emissions in developing countries are the major impediment to effective international climate policy, allocating responsibility for some portion of these emissions to final consumers elsewhere may represent an opportunity for compromise' (Davis & Caldeira, 2010, p. 5690). In other words, a consumption-based PPP may facilitate negotiations by ensuring that the poor will not be punished for production, and that the rich will shoulder the greatest burdens without being held liable for historical emissions—something they have dearly attempted to avoid.

4.5 Summary

In short, a present-oriented, consumption-based PPP better satisfies the convergence view that the rich ought to pay for climate action, and is more environmentally effective. It is also more sensitive to changing economic fates and may prove more politically feasible than alternatives. However, it is not free from problems. In what follows, I examine four potential objections concerning the applicability of the principle and its effects on the global poor. To avoid confusion with alternative PPPs, hereafter the present-oriented, consumption-based PPP is referred to as the *consumer-pays principle*, or CPP.

5   Objections

The first two objections elaborate on a concern expressed above: that present-oriented principles, like the CPP, *unduly* burden poor and developing states. I reject this concern in one form but accept it in another, which leads me to endorse a qualifying principle that limits the CPP's application and provides protections for economically disadvantaged states.

The latter two objections charge that the CPP is incomplete. One of these charges holds, and I respond by supplementing the CPP with an ability-to-pay principle (APP). The end result of these modifications is a pluralist, bi-phasic account of international climate justice, which is outlined below.

5.1   Unfair to the Global Poor (i): The CPP Does not Secure 'just Entitlements' to Emit

A popular view—defended (in various forms) by Baer, Neumayer, Caney, Jamieson, Singer, and others—is that everyone has an equal 'right' to generate a certain amount of emissions,[50] with quotas defined by the total absorptive capacity of the climate system divided by the number of people alive today (Caney, 2012; Gardiner, 2004, p. 583). On this view, the global rich, who have exceeded their quota, should

[50]See Gardiner (2004, p. 583ff), Caney (2005, p. 770; 2012), Neumayer (2000, pp. 185–192), Athanasiou and Baer (2002, pp. 76–97), Agarwal and Narain (1991), Jamieson (2001), Singer (2002, pp. 39–40), and Baer (2002). Politically, this view has been advocated by China, India and many less developed states.

pay for climate action, while the global poor, who are 'in credit,'[51] are entitled to further emissions (or owed compensation). Those taking this view might charge that, insofar as the CPP lacks a theory of just entitlements, it deprives the developing world of its 'fair share' of the atmospheric commons.

Although the intuitions behind claims about equal rights to emit appear generally sound, they become problematic in the particular context of climate change.[52] Given the extent to which climate change has already progressed, it is difficult, in consequentialist terms at least, to distinguish a right to emit from a right to harm. Simply too many emissions have been generated for fossil-fuel-based development to continue safely. Indeed, global mean temperatures will continue to increase as the total stock of GHGs in the atmosphere increases. Already atmospheric CO_2-e concentration levels have reached unprecedented and dangerous levels—as of February 2018, atmospheric CO_2-e exceeded 407 ppm, well above the commonly cited 'safe' upper limit of 350 ppm.[53] All new emissions beyond the earth's natural sequestration capacity can only compound the harmful effects of climate change.

One might respond here by emphasizing the relationship between emissions and standard-of-living—that is, by insisting that people must emit to sustain a minimally satisfactory life. Shue, for instance, defends a right to 'subsistence emissions' (Shue, 1993)—characterized as emissions necessary for securing a person's 'vital interests' (Shue, 1999, p. 541)[54]—and argues that this right places a duty on the rich to reduce their 'luxury emissions.' If this duty is observed, Shue and others argue, poor states could emit without jeopardizing current or future generations.

This response contains a fallacy, however: It conflates the right to a certain quality of life with a right to emit. Of course, until very

---

[51]Because 'their cumulative emissions are smaller than the global average per capita absorption' (IPCC, 1995, p. 94; quoted in Gardiner, 2004, p. 584).

[52]For additional criticisms, see Gardiner (2004, p. 583ff).

[53]Notably, according to ice-core samples, levels never exceeded 300 ppm in at least the last 800,000 years, and before the Industrial Revolution, CO_2e concentrations were around 280 ppm (Lüthi et al., 2008).

[54]Caney (2005) also adopts a vital-interest argument in defending a right to emit.

recently, emissions and standard-of-living were tightly correlated, such that increases in emissions were necessary to generate economic gains and thereby improve aggregate welfare. This relationship is not as rigid today as breakthroughs in renewable energy have made affordable, carbon-neutral development possible (Delucchi & Jacobson, 2011). To be sure, fossil-fuel industrialization may (in some cases) still offer the most expedient or inexpensive means for realizing welfare gains, but the desire to secure a certain standard-of-living does not justify using any means available. Moral prohibitions against harming must also be respected. That other states achieved development through fossil-fuel industrialization does not alter this, as merely citing the wrongdoing of others is insufficient for establishing standards of right or fairness.[55]

We should therefore reject claims about a right to emit founded on the right to a decent standard-of-living. Emitting GHGs in excess of the earth's absorptive capacity is harmful, and activities that unnecessarily[56] cause emissions should be limited if not prohibited.

5.2   Unfair to the Global Poor (ii): The CPP Is Not Poverty Sensitive

One might concede that a right to emit is problematic but still argue that the CPP is insensitive to developing states' interests—or sensitive only in a contingent way. For while a focus on consumption emissions *tends* to place the biggest burdens on the wealthiest states, should a poor country have high consumption emissions, as some (albeit not many) do, the CPP will impose correspondingly large burdens. Such burdens would jeopardize some states' ability to realize or maintain a decent standard-of-living (Caney, 2010, p. 213; see also Shue, 1999, 542).

---

[55]This would be a *tu quoque* fallacy.

[56]Unnecessarily can be interpreted in two ways here: (1) activities that contribute to climate change but are unnecessary for a satisfactory life (such as eating carbon-intensive foods, like meat and dairy products, when a plant-based diet is nutritionally sufficient and widely available) and (2) activities that are necessary for a satisfactory life but are undertaken in ways that unnecessarily result in greenhouse gas emissions (such as producing energy with coal or fracked gas when effective and clean alternatives like solar and wind are available).

The CPP should thus stipulate that climate-action burdens shall not undermine any state's ability to secure decent standard-of-living for its citizens.[57]

A few things to note about this proposed revision. First, while many argue that environmental-justice should not neglect broader distributive-justice concerns,[58] we must consider the possibility that combating climate change may not always comport with addressing economic injustice.[59] It is also possible that, given climate change's catastrophic potential, intergenerational justice may require deprioritizing today's global poor for the sake of future generations. Matthew Rendall (2011, pp. 891, 885), for instance, argues that while 'policies that deprived the poor of necessities so that the rich could continue their 'luxury emissions' ... would be a crying injustice,' this 'would be a lesser injustice than risking long-term catastrophe,' because 'the prospect of condemning several more generations in the South to poverty—terrible in itself—dwindles next to the danger of *permanent* impoverishment.'[60] In other words, so long as unabated climate change threatens the essential interests of future generations in a more intense or enduring way than poverty affects today's global poor, justice may require us to prioritize addressing the former ahead of the latter.

Yet, accepting this does not alter the basic intuition that allocating climate-action burdens in a way that *unnecessarily* undermines sufficient standard-of-living is unjust. Fortunately, there appears to be no strong reason to assume that responding to climate change must come at the cost of intra-generational economic justice (Gardiner, 2006; Rendall 2011; Singer, 2010, pp. 186, 55). The thrust of the objection must still be answered, then: Ideally, fair allocations of climate burdens

---

[57]I set aside the question of what counts as a sufficient minimum. Shue (1999, p. 541) defines it as 'enough for a decent chance for a reasonably healthy and active life of more or less normal length,' which extends beyond bare survival to those goods necessary for 'a distinctively human, if modest, life.' For other thresholds, see Shue (1993), Caney (2005, 2010, p. 218), and Singer (2002).

[58]Shue (1999), Caney (2005, 2012, pp. 258–259). But cf. Posner and Weisbach (2010).

[59]Posner and Weisbach (2010) stress this. I also take up this point in Mittiga (2018).

[60]Rendall (2011) later argues that imposing costs on poor states is probably unnecessary, assuming it is possible to shift the burdens of climate change onto future generations.

should strictly track contribution to the problem *and* be sensitive to capacity.[61] The CPP succeeds on the first front, but not necessarily on the second because it *tends* to impose greater burdens on rich states but is not constitutionally committed to this.

To meet this objection, we must recognize a qualifying principle external to the CPP. This principle presupposes a distinction between climate action itself and the costs of that action. For example, if the CPP results in the imposition of a universal carbon tax, this qualifying principle might hold that, at regular intervals, more advantaged states have a duty to provide tax rebates to less advantaged states. We can call this principle the economic justice qualifying principle (or EJQP). It can be expressed alongside the CPP as follows:

CPP: Climate-action burdens should be allocated in proportion to contribution, measured in terms of each state's annual consumption emissions.

EJQP: However, wealthy states[62] have a duty, in proportion to their wealth, to ensure that climate-action costs do not unnecessarily compromise any state's ability to attain or preserve decent standard-of-living.

One need not accept a thick cosmopolitan ethic to endorse the EJQP; rather, one need only to maintain that it is wrong to harm others, wherever they are in the world.[63] The IPCC's *Third Assessment Report* stresses that '[a]ny individuals' or nations' actions to address the climate-change issue, even the largest emitting nation acting alone, can have only a small effect.'[64] In other words, each state requires the cooperation of all or most others to mitigate climate change successfully. Consequently, attempting to structure international climate action in a way that

---

[61]This idea is reflected in the 'common but differentiated responsibilities' doctrine (UNFCCC, 1).

[62]In this context, 'wealthy' refers to states belonging to the World Bank's 'high-income' and 'upper middle-income' groups. For an inventory of these states, and of those in the 'lower middle' and 'low-income' groups, see IPCC, AR5, WG3, A.II.2.3, pp. 1287–1288.

[63]The argument in this paragraph draws on Shue (1999, p. 541ff).

[64]IPCC, AR3, WG3, 607. See also IPCC, AR5, WG3, pp. 5, 214, 136.

unnecessarily undermines the realization or preservation of a decent minimum standard-of-living in developing states amounts to a will to harm—provided, as Shue notes, 'that interfering with people's ability to maintain a minimum for themselves count[s] as a serious harm' (Shue, 1999, p. 542). Put more simply, if forcing the least advantaged to sacrifice is unnecessary, it is also harmful and unfair, regardless of other considerations.

Note, however, that this does not relieve poor and developing states of their climate-*action* responsibilities—it does not imply, for instance, that poor states are entitled to delay mitigation policies. Poor and developing states, like all states, have duties not to exacerbate climate change, but satisfying this duty can rightly be predicated on more advantaged states fulfilling their obligation, specified by the EJQP, to ensure that the strains of international cooperation do not unnecessarily compromise the vital interests of the least well off.

5.3   Incomplete (i): The CPP Cannot Ground Duties to Enhance Carbon Sinks

Simon Caney distinguishes between 'atomist' and 'holist' accounts of climate justice. Atomist accounts offer a separate and distinct principle for each climate burden (such as mitigation, adaptation, and compensation), whereas holist accounts treat climate burdens 'en masse,' with a single principle (Caney, 2012, pp. 258–259).[65] I have presented my account as holist. One might challenge this, however, on the grounds that the CPP is fundamentally aimed at discouraging bads (such as the generation of greenhouse gas emissions), not promoting goods, and thus cannot provide for the *enhancement* of carbon sinks, like forests and certain marine habitats (Armstrong, 2016; Duarte et al., 2013; Page, 2016, p. 85). Thus, for any account to be truly holist, it must be able to offer a principled basis for ensuring that sinks are properly maintained and duly expanded.

---

[65]Caney also notes that there can be intermediate accounts covering some but not all climate burdens.

Although this is a serious objection, a simple response may be available. With estimates of the annual sequestration capacity of particular forests and marine habitats, credits could be awarded to the states maintaining them. This would simply require regarding the sinks as consumable goods that provide annual returns (in the form of carbon sequestration capacity)—a kind of rent-deriving property.

Several advantages would follow from this. First, providing credits for sinks would open a stream of benefits for poor and developing states, thereby correcting for the disproportionate burdens they currently bear in preserving what are, after all, collective goods (Armstrong, 2016; Page, 2016, p. 89). Relatedly, if an international market was established in which the rights to these credits could be leased, poor and developing states could secure direct financial transfers from wealthy states seeking to lower their yearly emissions totals, without ceding control of the territories hosting the sinks. Second, awarding credits would incentivize the maintenance and expansion of carbon sinks. Indeed, a credit system effectively doubles the value of a sink since razing a forest would entail both losing a credit (equivalent to the sequestration capacity) and incurring a fee (equivalent to the carbon emitted from the land-use change). These calculations may seem complex, but factoring sequestration credits and land-use changes into national emissions estimates is already common practice; thus, incorporating them into a consumption-based model poses no insuperable difficulties.

5.4   Incomplete (ii): The CPP Cannot Allocate Burdens Without Human Pollution

The final objection is that the CPP cannot provide a coherent basis for allocating climate-action burdens in two important cases: when human activity (i) is not or (ii) is no longer driving climate change.

5.4.1 Anthropogenic and Non-anthropogenic Climate Change

Consider, first, the IPCC's claim (cited in Caney, 2010, p. 211) that 'most of the warming observed over the last 50 years is attributable to human activities.' As this implies, other natural processes also contribute to global warming, if in a far less pronounced way. This poses a problem for the CPP. Specifically, because the CPP allocates burdens in proportion to contribution, distinguishing between anthropogenic and non-anthropogenic climate change seems necessary. In making this distinction, however, another problem arises: The CPP appears unable to address non-anthropogenic climate change.[66]

In response, we should first note that non-anthropogenic climate change would almost certainly not be a cause for concern were it not for our gross exacerbation of the problem (US Environmental Protection Agency). Given this, polluters should be held responsible for the problem as a whole. We know (and have known for decades) that climate change is real and that human activity is causing changes that almost certainly would not have occurred without our interference. In this sense, when we contribute to climate change, in awareness of what our actions entail, this confers on us a general responsibility for the outcomes that follow, *even if* the problem might have occurred, to some extent, without our interference.

Moreover, by dint of being present oriented, the CPP has a more expansive notion of contribution—one that includes damages caused *and* risk imposed. That is, when internalizing an activity's social costs, the CPP includes a 'risk premium,' which reflects the magnitude and likelihood that damages or losses associated with that activity will come to pass. Notably, such risks need not be caused *exclusively* by human activity. If a given risk is great enough, society may simply wish to ensure that all activities contributing to it are discouraged or stopped. Consequently, distinguishing between anthropogenic and non-anthropogenic climate change is unnecessary. A reasonable aversion to risk in general, and the knowledge that human activities are increasing the likelihood or potential magnitude of a given risk, suffices for grounding responsibility.

---

[66]Caney (2010, p. 211) registers a version of these concerns against his backward-looking PPP.

5.4.2 Climate Change Without Polluters

The second charge—that the CPP cannot allocate shares of responsibility once human activity ceases to contribute to climate change—is more difficult to address. Consider the following. If the CPP is successful, emission flows will decline, perhaps falling below the earth's sequestration capacity before long. Yet, even with an immediate and precipitous emissions drop, climate change may continue to cause problems for centuries to come. This, again, is because many GHGs endure in the atmosphere long after they are emitted.[67] Consequently, distributing the costs of adaptation and compensation will likely remain an important international issue well after we reach the point of carbon neutrality. Given that the CPP allocates duties in proportion to present contribution, however, it seems inapplicable during a 'post-mitigation' period. As emissions decline, eventually the revenue the CPP generates from justly priced taxes on GHGs will be insufficient for covering the expenses related to adaptation and compensation. Thus, the CPP is incomplete.

Answering this objection requires supplementing the CPP with a principle that can explain how to correct for any deficits between the revenue generated by taxing emissions and the total cost of climate-action burdens for any given year. There are two clear possibilities: a BPP and an APP.

Recall that a BPP assigns burdens to those who have benefitted from the activities that gave rise to climate change, in proportion to their benefit (Page, 2016). While this seems plausible within intermediate time horizons, it becomes incoherent when applied to the distant future. Recall that climate change will likely persist far after the point that dangerous emissions are generated. What would a BPP commit us to? Would it be fair, 500 years from today, to hold a completely carbon-neutral country responsible for the remaining burdens of climate action because of the benefits its citizens once received from fossil-fuel

---

[67]For instance, while about 60% of carbon dioxide (CO_2)—the most common GHG—will cycle out of the atmosphere within 200 years after being released, up to 20% will remain for 'tens of thousands of years' (Hausfather, 2010).

industrialization? Would it matter if economic fates shift over this time—if a once-rich country becomes relatively poor, for instance? What if it is no longer a state at all? As time progresses, these questions compound, making the BPP less and less coherent.

We might attempt to preempt these issues by isolating the stream of wealth directly generated by greenhouse gas emissions and using this as the limit of liability. Page, for instance, argues that responsibility under the BPP ought to end at the point that the 'benefits traceable to activities that drive climate change are exhausted' (Page, 2016, p. 91). Isolating the particular benefits derived from climate-inducing activities would be a tremendous practical challenge.[68] Assuming this could be done, however, we might wonder what to do when the limit is reached. Given the long atmospheric lives of many GHGs, it is possible that this stream of wealth will be exhausted well before adaptation and compensation are no longer concerns. If this is right, then the BPP will itself have to be supplemented, thus raising again the original problem.

A more parsimonious—and less theoretically fraught—solution would be to supplement the CPP with the APP, which again holds that the wealthy should pay proportionately for the costs of climate action. Darrell Moellendorf, Caney, Page, and others have used the APP to supplement the central principles in their accounts (typically a backward-looking PPP or BPP) (Caney, 2005, 2010; Moellendorf, 2002, pp. 97–100; Page, 2011; Shue, 1999). This seems appropriate here as well. In other words, in a post-mitigation phase of climate change, the APP likely offers the most coherent basis for allocating climate duties.

6   Conclusion

We now have a pluralist, bi-phasic account of climate justice, the three pillars of which are as follows:

---

[68]How can we isolate a benefit that arose from activities that cause climate change from those resulting from, e.g. sea access, education investments, or luck? An agent's economic success is predicated on numerous factors, a mere inventory of which would be confounding.

CPP: Climate-action burdens should be allocated in proportion to contribution, measured in terms of each state's annual consumption emissions.

EJQP: However, wealthy states have a duty, in proportion to their wealth, to ensure that climate-action costs do not unnecessarily compromise any state's ability to attain or preserve a decent standard-of-living.

APP: Once consumption emissions decrease to the point that the revenue gained from taxing them can no longer sustain the remaining costs (related to adaptation and compensation claims), wealthy states should shoulder those burdens in proportion to their wealth.

While this account is not as picturesque as one that simply holds 'the polluter should pay,' it is markedly more coherent; and, because it covers all the relevant climate burdens, now and into the future, it is also more comprehensive. Moreover, the account is alive to both the contributions and capacities of different actors in both phases. In particular, the first phase of the account, covered by the CPP and EJQP, is contribution determined and capacity sensitive, while the latter phase, covered by the APP, is determined by and sensitive to capacity. In this regard, it is responsive to the claims of both compensatory and distributive justice.

Several questions remain. For instance, should historical emissions after the excusable ignorance cut-off date be taken into account and, if so, how? Should the CPP apply at the subnational level? And to what extent, if any, should we discount future welfare when setting emissions tax rates? These questions must be addressed in future work. The aim here, however, has been simply to show that the PPP (*qua* CPP) can provide a politically feasible, ethically compelling, and environmentally effective basis for allocating climate burdens among states.

References

Agarwal, A., & Narain, S. (1991). *Global warming in an unequal world: A case of*. New Delhi: Centre for Science and Environment.

Armstrong, C. (2016). Fairness, free-riding and rainforest protection. *Political Theory, 44*(1), 106–130.

Arrow, K. J., Parikh, J., & Pillet, G. (1995). *Decision-making frameworks for addressing climate change* (IPCC Second Assessment Report). IPCC.

Athanasiou, T., & Baer, P. (2002). *Dead heat: Global justice and global warming.* New York: Seven Stories Press.

Baer, P. (2002). Equity, greenhouse gas emissions, and global common resources. In S. H. Schneider, A. Rosencranz, & J. O. Niles (Eds.), *Climate change policy: A survey* (pp. 393–408). Washington, DC: Island Press.

Baer, P. (2010). Adaptation to climate change: Who pays whom? In S. Gardiner, S. Caney, D. Jamieson, & H. Shue (Eds.), *Climate ethics: Essential readings* (pp. 247–262). New York, NY: Oxford University Press.

Barrett, S. (1998). Political economy of the Kyoto Protocol. *Oxford Review of Economic Policy, 14*(4), 20–39.

Bell, D. (2011). Does anthropogenic climate change violate human rights? *Critical Review of International Social and Political Philosophy, 14*, 99–124.

Brooks, T. (2015). Climate change justice through taxation? *Climatic Change, 133*, 419–426.

Broome, J. (2012). *Climate matters: Ethics in a warming world.* New York: Norton.

CAIT Climate Data Explorer. (n.d.). Washington, DC: World Resources Institute. Retrieved August 9, 2016, from http://cait.wri.org.

Caney, S. (2005). Cosmopolitan justice, responsibility, and global climate change. *Leiden Journal of International Law, 18*(4), 747–775.

Caney, S. (2006). Environmental degradation, reparations and the moral significance of history. *Journal of Social Philosophy, 37*(3), 464–482.

Caney, S. (2009). Climate change and the future: Discounting for time, wealth, and risk. *Journal of Social Philosophy, 40*(2), 163–186.

Caney, S. (2010). Climate change and the duties of the advantaged. *Critical Review of International Social and Political Philosophy, 13*(1), 203–228.

Caney, S. (2012). Just emissions. *Philosophy & Public Affairs, 40*(4), 255–300.

Davis, S. J., & Caldeira, K. (2010). Consumption-based accounting of CO_2 emissions. *Proceedings of the National Academy of Science, 107*(12), 5687–5692.

Delucchi, M. A., & Jacobson, M. Z. (2011). Providing all global energy with wind, water, and solar power, Part II: Reliability, system and transmission costs, and policies. *Energy Policy, 39*, 1170–1190.

Droege, S. (2011). Using border measures to address carbon flows. *Climate Policy*, 1191–1201.

Duarte, C. M., Losada, I. J., Hendriks, I. E., Mazarrasa, I., & Marbà, N. (2013). The role of coastal plant communities for climate change mitigation and adaptation. *Nature Climate Change, 3,* 961–968. https://doi.org/10.1038/nclimate1970.

Eckersley, R. (2015). The common but differentiated responsibilities of states to assist and receive 'climate refugees'. *European Journal of Political Theory, 14*(4), 481–500.

Furman, J., Shadbegian, R., & Stock, J. (2014). *The cost of delaying action to stem climate change.* Council of Economic Advisers. Washington, DC: White House. Retrieved March 3, 2015, from https://www.whitehouse.gov/sites/default/files/docs/the_cost_of_delaying_action_to_stem_climate_change.pdf.

Gardiner, S. M. (2004). Ethics and global climate change. *Ethics, 114*(3), 555–600.

Gardiner, S. M. (2006). A core precautionary principle. *The Journal of Political Philosophy, 14*(1), 33–60.

Gardiner, S. M. (2011). *A perfect moral storm: The ethical tragedy of climate change.* Oxford: Oxford University Press.

Gosseries, A. (2004). Historical emissions and free-riding. *Ethical Perspectives, 11*(1), 36–60.

Hausfather, Z. (2010, December 12). *Common climate misconceptions: Atmospheric carbon dioxide.* Retrieved August 9, 2016, from Yale Climate Connections http://www.yaleclimateconnections.org/2010/12/common-climate-misconceptions-atmospheric-carbon-dioxide/.

Huq, H., Roberts, E., & Fenton, A. (2013). Loss and damage. *Nature Climate Change, 3,* 949.

India's Intended Nationally Determined Contribution: Working Towards Climate Justice. (2015). UNFCCC. Retrieved from http://www4.unfccc.int/ndcregistry/PublishedDocuments/India%20First/INDIA%20INDC%20TO%20UNFCCC.pdf.

James, R., Otto, F., Parker, H., Boyd, E., Cornforth, R., Mitchell, D., & Allen, M. (2014). Characterizing loss and damage from climate change. *Nature Climate Change, 4,* 938–939.

Jamieson, D. (2001). Climate change and global environmental justice. In P. Edwards & C. Miller (Eds.), *Changing the atmosphere: Expert knowledge and global environmental governance* (pp. 287–307). Cambridge: MIT Press.

Jamieson, D. (2010). Adaptation, mitigation, and justice. In S. M. Gardiner, S. Caney, D. Jamieson, & H. Shue (Eds.), *Climate ethics: Essential readings* (pp. 263–283). New York: Oxford University Press.

Joint Science Academies' Statement: Global Response to Climate Change. (2005). Retrieved from http://www.royalsoc.ac.uk/displaypagedoc.asp?id=20742.

Kander, A., Jiborn, M., Moran, D. D., & Wiedmann, T. O. (2015). National greenhouse-gas accounting for effective climate policy on international trade. *Nature Climate Change, 5,* 431–435.

Kingston, E. (2014). Climate justice and temporally remote emissions. *Social Theory and Practice, 40*(2), 281–303.

Lenzen, M., Kanemoto, K., Moran, D., & Geschke, A. (2013). Building eora: A global multi-regional input-output database at high country and sector resolution. *Economic Systems Research, 25*(1), 20–49.

Lowe, J. A., Gregory, J. M., Ridley, J., Huybrechts, P., Nicholls, R. J., & Collins, M. (2006). The role of sea-level rise and the Greenland ice sheet in dangerous climate change: Implications for the stabilisation of climate. In H. J. Schellnhuber, W. Cramer, N. Nakicenovic, T. Wigley, & G. Yohe (Eds.), *Avoiding dangerous climate change.* Cambridge: Cambridge University Press.

Lüthi, D. (2008). IGBP PAGES/World data center for paleoclimatology data contribution series # 2008-055. *EPICA Dome C Ice Core 800KYr Carbon Dioxide Data.* Boulder, CO: NOAA/NCDC Paleoclimatology Program. Retrieved from http://cdiac.ornl.gov/trends/co2/ice_core_co2.html.

Lüthi, D., Floch, M. L., Bereiter, B., Blunier, T., Barnola, J.-M., et al. (2008). High-resolution carbon dioxide concentration record 650,000-800,000 years before present. *Nature, 453,* 379–382.

MacCleery, D. W. (2011). *American forests: A history of resiliency and recovery.* Durham, NC: Forest History Society. Retrieved from http://foresthistory. org/Publications/Issues/American_Forests.pdf.

Miller, D. (2009). Global justice and climate change: How should responsibilities be distributed? *The Tanner Lectures on Human Values,* 119–156. Retrieved from http://tannerlectures.utah.edu/_documents/a-to-z/m/Miller_08.pdf.

Mittiga, R. (2018). *Before collapse: A political theory of climate catastrophe.* Charlottesville, VA: Unpublished dissertation.

Moellendorf, D. (2002). *Cosmopolitan justice.* Boulder, CO: Westview Press.

Moellendorf, D. (2012). Climate change and global justice. *Wiley Interdisciplinary Reviews: Climate Change, 3,* 131–143.

National Oceanic and Atmospheric Administration (NOAA); U.S. Department of Commerce; National Ocean Service; Center for Operational Oceanographic Products and Services. (2017). Global and regional sea level rise scenarios for the United States. Retrieved from https://tidesandcurrents. noaa.gov/publications/techrpt83_Global_and_Regional_SLR_Scenarios_ for_the_US_final.pdf.

Neumayer, E. (2000). In defence of historical accountability for greenhouse gas emissions. *Ecological Economics, 33,* 185–192.

Nordhaus, W. D. (2009). *A question of balance: Weighing the options on global warming policies.* New Haven, CT: Yale University Press.

Organisation for Economic Co-operation and Development. (1995). *Environmental principles and concepts.* Paris. Retrieved from http://www. oecd.org/trade/envtrade/39918312.pdf.

Page, E. A. (2011). Climatic justice and the fair distribution of atmospheric burdens: A conjunctive account. *The Monist, 94*(3), 412–432.

Page, E. A. (2016). Qui bono? Justice in the distribution of the benefits and burdens of avoided deforestation. *Res Publica, 22,* 83–97.

Peters, G., & Hertwich, E. (2008a). CO_2 embodied in international trade with implications for global climate policy. *Environmental Science and Technology, 42*(5), 1401–1407.

Peters, G., & Hertwich, E. (2008b). Post-Kyoto greenhouse gas inventories: Production versus consumption. *Climate Change, 86,* 51–66.

Peters, G. P., Minx, J. C., Weber, C. L., & Edenhofer, O. (May 2011). Growth in emission transfers via international trade from 1990 to 2008. *Proceedings of the National Academy of Sciences, 108*(21), 8903–8908. https://doi. org/10.1073/pnas.1006388108.

Posner, E. A., & Weisbach, D. (2010). *Climate change justice.* Princeton, NJ: Princeton University Press.

Rawls, J. (1999). *A theory of justice: Revised edition.* Cambridge, MA: Harvard University Press.

Rendall, M. (2011). Climate change and the threat of disaster: The moral case for taking out insurance at our grandchildren's expense. *Political Studies, 59,* 884–899.

Schüssler, R. (2011). Climate justice: A question of historic responsibility? *Journal of Global Ethics, 7,* 261–278.

Scientific American. (n.d.). *Is the world outsourcing its greenhouse emissions to China?* Retrieved April 29, 2018, from https://www.scientificamerican.com/ article/earth-talks-outsourcing-greenhouse-china/.

Senate, U. (1997, July 25). S. Res. 98—A resolution expressing the sense of the Senate regarding the conditions for the United States becoming a signatory to any international agreement on greenhouse gas emissions under the United Nations framework convention on climate change. *Legislation.* 105th Congress (1997–1998). Retrieved from https://www.congress.gov/ bill/105th-congress/senate-resolution/98/text.

Shue, H. (1993). Subsistence emissions and luxury emissions. *Law & Policy, 15*(1), 39–59.

Shue, H. (1999). Global environment and international inequality. *International Affairs, 75*(3), 531–545.

Singer, P. (2002). *One world: The ethics of globalization.* New Haven, CT: Yale University Press.

Singer, P. (2010). One atmosphere. In S. M. Gardiner, S. Caney, D. Jamieson, & H. Shue (Eds.), *Climate ethics: Essential readings* (pp. 181–199). Oxford: Oxford University Press.

Steinfeld, H., Gerber, P., Wassenaar, T., Castel, V., Rosales, M., & De Haan, C. (2006). Livestock's Long Shadow: Environmental Issues and Options. Food and Agriculture Organization of the United Nations, Rome. Retrieved from ftp://ftp.fao.org/docrep/fao/010/a0701e/a0701e.pdf.

Steininger, K., Lininger, C., Droege, S., Roser, D., Tomlinson, L., & Meyer, L. (2014). Justice and cost effectiveness of consumption-based versus production-based approaches in the case of unilateral climate policies. *Global Environmental Change, 24,* 75–87.

Steininger, K., Lininger, C., Meyer, L., Munoz, P., & Schinko, T. (2016). Multiple carbon accounting to support just and effective climate policies. *Nature Climate Change, 6,* 35–41.

Stern, N. (2007). *The Stern review: The economics of climate change.* Cambridge: Cambridge University Press.

Stern, N. (2010). The economics of climate change. In S. M. Gardiner, S. Caney, D. Jamieson, & H. Shue (Eds.), *Climate ethics: Essential readings* (pp. 39–76). Oxford: Oxford University Press.

Trump, D. (2017, June 1). *Statement by President Trump on the Paris climate accord.* Retrieved April 2018, 29, from https://www.whitehouse.gov/briefings-statements/statement-president-trump-paris-climate-accord/.

United Nations Framework Convention on Climate Change. (1992). Paris. Retrieved from https://unfccc.int/resource/docs/convkp/conveng.pdf.

United States Environmental Protection Agency. (n.d.). *Causes of climate change.* Retrieved September 26, 2016, from https://www3.epa.gov/climatechange/science/causes.html.

Wagner, G., & Weitzman, M. (2016). *Climate shock: The economic consequences of a hotter planet.* Princeton, NJ: Princeton University Press.

Weitzman, M. (2007). A review of the Stern review on the economics of climate change. *Journal of Economic Literature, 45*(3), 703–724.

Wellesley, L., Happer, C., & Froggatt, A. (2015). *Changing climate, changing diets: Pathways to lower meat consumption.* Chatham House: The Royal

Institute of International Affairs, London, UK. Retrieved June 1, 2016, from https://www.chathamhouse.org/sites/files/chathamhouse/publications/research/CHHJ3820%20Diet%20and%20climate%20change%2018.11.15_WEB_NEW.pdf.

Whiteman, G., Hope, C., & Wadhams, P. (2013). Climate science: Vast costs of Arctic change. *Nature, 499.*

9

Comparison of Human and Non-human Migration Governance Under Climate Change

Martina Grecequet, Jessica J. Hellmann,
Jack DeWaard and Yudi Li

1   Introduction

Climate change is expected to affect the migration of people worldwide, such that 140 million people are projected to migrate within their own states by 2050 (Rigaud et al., 2018). Others estimate that about 1.6 billion people will be displaced due to sea level rise by 2060 (Geisler & Currens, 2017). The relationship between climate and human migration depends on the vulnerability of people and places to climate change. Technological innovations especially in developed states have reduced the vulnerability of some people to climate variation and thus allowed humans to settle in areas with fewer or less reliable resources. Yet, there are about 380 million smallholder farms around the world

M. Grecequet (✉) · J. J. Hellmann · J. DeWaard · Y. Li
University of Minnesota, Minneapolis, MN, USA
e-mail: mgrecequ@umn.edu

J. J. Hellmann
e-mail: hellmann@umn.edu

© The Author(s) 2019   **195**
B. Edmondson and S. Levy (eds.), *Transformative Climates and Accountable Governance*, Palgrave Studies in Environmental Transformation, Transition and Accountability, https://doi.org/10.1007/978-3-319-97400-2_9

who depend on favourable climates, and these people have climate change risks that could drive new waves of migration (Samberg, Gerber, Ramakutty, Herrero, & West, 2016).

Environmental changes brought on by climate change also affect the distribution of global biodiversity, on which humans critically depend. The redistributions of species are accelerating, and there is also evidence of changes in species composition and abundance decline due to climate change (CBD, 2017). The extinction of many species continues at a rate thousands of times faster than natural background rates (De Vos, Joppa, Gittleman, Stephen, & Pimm, 2015). As temperatures rise, the survival of some plant and animal species will depend on their abilities to adapt to changes in their historical habitat or to migrate to new habitats with more favourable climate conditions (Pelc et al., 2017; Ryan et al., 2018). Studies suggest that if they fail to migrate hundreds of thousands of species could become extinct (Javeline et al., 2015).

The migrations of people have been studied primarily in isolation from the migrations of other species, to the extent that these now typically constitute two separate research domains. However, the general theory of human migration was developed from the theory and laws that were discovered from studying the migrations of different plant and animal species (Ratzel, 1882; Ravenstein, 1885; Wagner, 1873). Thus, there are important similarities. Both the biological literature and social science literature recognize that migration can be an adaptation strategy because it offers new opportunities for group survival in the face of changing conditions. The ability to migrate is, however, not evenly spread among people or species. For example, studies suggest that there are many natural (such as physiological) and social (such as income) barriers to migration (Beever et al., 2016; Black, Bennett, Thomas, & Beddington, 2011). Thus, some people and species could be trapped in vulnerable regions, reducing the efficiency of migration as an adaptation strategy.

In this chapter, we explore key similarities and differences between human and non-human migration due to climate change drawing upon ecological and social science perspectives. We argue this comparison has value because knowledge from one field can shed light on the need for research and action in the other. Ecologists have long-studied the movements, evolutions and extinctions of species as responses to climate and other environmental changes, while social scientists have studied migration processes in the context of cultural, economic, environmental, political and social factors (Lee, 1966; Zelinsky, 1971). Until recently, environmental- and climate-related factors were perceived as having negligible influences on human migration patterns (McLeman, 2014). However, recent studies show that environmental factors are likely to be important drivers of human migration in the coming decades (Black et al., 2011). As a global society, we currently have greater than ever need to understand, predict and manage migratory processes for the benefit of human and non-human species. Different traditions in human and non-human demography can cross-pollinate for the benefits of planners, regulators and other decision-makers.

Meeting the challenges of climate change through effective management of migration requires new legal and policy approaches that address movements across local, national and international boundaries. In addition, human and non-human migration also affect each other both directly and indirectly through changes in the distribution of goods and services. The policies that govern human migration may also affect the movements of animal and plant species and vice versa, such as climate adaptation or resiliency projects for people, or the placement of parks and reserves for species conservation. Overall, this chapter is dedicated to promoting shared learning, with the goal of accelerating beneficial practices for humans and non-humans alike.

2   Migration as Adaptation Strategy

We might need to see migration within a pattern of a longer search for the good, for survival, resilience and flourishing. This challenges the way that we might think theologically, politically, economically and ecologically about migration.

Anna Rowlands (2018)

Different people have different vulnerabilities to climate change, and we must confront the moral and practical implications of these differences. The same is true for plants and animals. We must identify effective adaptation strategies to help people and species to deal with the negative effects of climate change, and we must apply those strategies to the most vulnerable among us. The discussion and application of adaptation are complicated by different uses of the term across research communities. In the biological sciences, adaptation usually refers to an evolutionary process: an increase in genetically based traits that enhance survival and reproductive success within populations (Darwin, 1859). In the climate and development community, however, adaptation refers to human adjustments to changing conditions (IPCC, 2014). Adjustments without evolution can also happen in non-human populations, and humans can help along both evolutionary and non-evolutionary adjustments.

Despite some vocabulary differences, there is much that the human and non-human climate change communities—those working to help systems adjust to change—can learn from one another. Here, we use 'adaptation' to refer to the broad set of actions that people can take to help both human and non-human systems survive, and possibly thrive, through changing climatic conditions. In *human systems*, adaptation strategies can also be highly technical, including, for example, building a seawall or constructing a heat-mitigating green roof. Human adaptation strategies can also involve social and cultural options such as changing agricultural practices or reforming disaster preparedness. In *natural systems*, adaptation options can also be quite technical such as wetland restoration or urban green space, as well as large-scale interventions that involve complex social and political systems such as reestablishing landscape connectivity, relocation of species (Richardson et al., 2009; Schwartz, Hellmann, & McLachlan, 2009) or harvesting or other management practices. There is considerable overlap of strategies between human and non-human adaptation, including migration/relocation, recreation of natural assets and shifts in policy or practice.

2.1   Migration

In *human systems*, migration is defined as the spatial movement of a person which requires a change in the place of usual residence and which involves crossing a recognized political or administrative border

(White, 2016). Demographers distinguish between two categories of migrants: internal migrants who move from a residence in one state/province or from on district/municipality/county; and international migrants who move to a new residence across a national boundary, from one state to another. Migration patterns differ from state to state. The decision to migrate is often a result of multiple factors, which might include a combination of economic, political, social, demographic and environmental factors, underpinning motivations for either a person or whole family to leave or stay in their current location. These motivation factors are closely interlinked, and it makes little sense to consider them in isolation (Black et al., 2011). Climate vulnerability, our predisposition to be negatively affected by climate change, underlies many of these factors, however, and affects the ability to migrate (Black et al., 2011; IPCC, 2012). There is consensus among migration scholars that most climate-related migration in the twenty-first century will be internal and of shorter distance than international migrations which involve long distances (Hunter & Nawrotzki, 2016). However, there is evidence of international migration also occurring due to environmental and climate factors, for example, from Pacific islands to New Zealand and Australia (Locke, 2009).

In *natural systems*, migration also entails complex processes, arising from interactions among genes, individuals, other species and the physical environment. Short-distance dispersal for some species is a natural component of reproduction. Except for migratory or highly mobile species, however, long-distance dispersal is relatively rare (Higgins & Richardson, 1999). Yet, long-distance dispersal drives large-scale processes such as the size of a species' geographic range (Nathan, 2006), and evidence suggests that migration was the dominant response of species to past climatic changes (such as during the Quaternary) (Davis & Shaw, 2001). Migration can also drive evolution if migrants have higher survival and reproduction rates than non-migratory groups (Davis & Shaw, 2001; Hill, Griffits, & Thomas, 2011), and migration can evolve as a strategy within species in environments with significant spatial and temporal variations in resources (Cresswell, Satterhwaite, & Sword, 2011).

Evolutionary change caused by migration in response to human-caused climate change has been observed already in some cases (e.g. longer jumping legs in grasshoppers). Knowing that migration was important historically leads ecologists to expect it under modern climate change, and ecologists are striving to understand which species both require and are better able to move in response to climate change. It is possible that seasonal migrant species—by nature of their ability to readily move—may fare better under a changing climate than resident species. Henceforth, we will refer to migration as a range shifts/extends across latitude and elevation, rather than seasonal flow (Clark et al., 1998). This use also aligns with a permanent or semi-permanent flow of people from one place of residence to another.

2.2 Defining Vulnerability

In both human and natural systems, the need for adaptation—whether migration or otherwise—is a function of vulnerability that includes four elements: occurrence of climate hazard (H), exposure to climate hazard (E), a sensitivity (S) and adaptive capacity (AC) within the system to confront, mitigate or otherwise reduce the consequences of exposure and sensitivity (IPCC, 2014) (see Table 1). The definition of vulnerability under climate change is similar for human and non-human systems (Williams, Shoo, Isaac, Hoffmann, & Langham, 2008), where the ability to migrate is an adaptive capacity. The potential for migration increases with the severity of the event, with the sensitivity of the exposed population and with lesser non-migratory adaptive capacity in both cases. When we manage, promote or enable migration in a policy, programme or another form of action, it is an adaptation strategy.

All living things—both human and non-human—are exposed to the same natural hazards caused by climate change (e.g. heat waves, increased precipitation and flooding), as well as hazards aggravated by other forms of environmental degradation (such as increased climate impacts on degraded lands).

Sensitivity is a degree to which people and other species are affected by climate variability. Sensitivity is different among species and among

Table 1 Definition of terminology related to climate vulnerability for human and natures systems. Based on AR5 IPCC=Intergovernmental Panel on Climate Change Fifth Assessment Report (2014), Cardona et al. (2012), and Williams et al. (2008)

Term	IPCC definition	Human system	Nature system
Hazard (H)	The potential occurrence of natural or human-induced physical event	Climate hazard may cause lose of life, injury, loss of property and infrastructure	Climate hazard causes loss of life, injuries, loss of environmental resources and habitat
Exposure (E)	A degree to which nature and people are exposed to climate hazard	Population and assets located in areas affected by hazards (e.g. flood zone of a city)	Governed by factors that are extrinsic to the species. Presence of species or ecosystems in places that could be adversely affected
Sensitivity (S)	Degree to which human or nature system is affected either adversely or beneficially	Determined by society dependency on climate-sensitive sectors (agriculture), physical impairment (elderly)	Determined by how tightly linked a species is to current conditions, their physiological tolerance limits, ecological traits and genetic diversity
Adaptive capacity (AC)	The ability of systems, humans and other organisms to adjust to potential damage to take advantage of opportunities or to respond to consequences	Potential to utilize social and economic resources to implement adaptation measures	Ecological and evolutionary responses, potentially modifying physiological tolerance limit, dispersal activity and behaviour adjustment

people, capturing, for example, human and non-human populations that are sensitive to climate hazards due to their living locations
(low-lying coastal areas). Another example is subsistence farmers that
are particularly dependent on rainfall amounts and pattern and thus
are more sensitive to climate variability than agriculture with irrigation
or non-agricultural economies. Other examples include species that do
not regulate their own body temperature—for instance most insects—
which are sensitive to ambient temperature because they move and
reproduce only when it falls in a particular range (Deutsch et al., 2008;
Wilson, Hustler, Ryan, Burger, & Noldeke, 1992). Physiology plays
an important role in the sensitivity of humans too. On average, the
elderly have low thermal tolerances and thus experience increased mor-
tality rates during heat waves (Stafoggia et al., 2006; Schwartz, 2005).
Together, exposure and sensitivity determine the level of risk that a system experiences under climate change.

Adaptive capacity counteracts climate risk. Adaptive capacities are
those behavioural, technological, evolutionary or other available adjustments that humans and non-humans can use to reduce their exposure
or sensitivity. In adaptive capacity, there are clear differences between
humans and non-humans, as the tools available to reduce risk are often
socio-economic in humans and biological in non-human systems. Still,
there are valuable parallels between the two, suggesting possible actions
that can be taken to reduce vulnerability. While some adaptation
actions reduce exposure or sensitivity, the vast majority of adaptation
lies in increasing the adaptive capacity of a system.

The social and technological innovations, such as irrigation or
drought resilient crops, which are used in agriculture, have in many
ways decoupled human dependency on local climate, at least in states
and regions that are less dependent on subsistence farming, fishing or
forestry, all of which might otherwise be considered climate-sensitive
sectors. The use of air conditioning or heating systems also helps
humans to regulate the temperatures they experience in less suitable
climates.

The adaptive capacity of non-human species includes 'phenotypic
plasticity' (changes in species behaviour, morphology and physiology
in response to changes in their environments), 'dispersal ability'

(the ability to move to more suitable microclimates within their historic range or area of occupancy or to move to new areas that were not historically occupied) and 'evolutionary change' (changes in the frequency of traits within a population or species as changing conditions change which traits are favourable) (Beever et al., 2016). There are limits to all of these capacities. The first two are limited by physiology, morphology and behaviour. The last is limited by the amount and type of genetic variation that exists within a species (Blows & Hoffmann, 2005; Hellmann & Pineda-Krch, 2007; Husby, Visser, & Kruuk, 2011). New variation is always emerging within natural populations due to mutation, but this process is generally slow relative to the rate of human-caused climate change (Hoffmann & Sgro, 2011).

The ability to adapt to changing conditions through migration is common to both humans and non-humans, with some people and species being better able to take advantage of migration more than others. In both human and non-human systems, the necessity for migration generally increases when local adaptive capacity is limited. Thus, migration can be seen as an alternative to in situ adjustment—hence, the non-human mantra for climate change: 'adjust, move or die'. In other words, extinctions of species can occur where in situ changes or migrations do not—or cannot—take place (Lenoir & Svenning, 2014).

3   Trapped by Climate

Human migration occurs for a variety of reasons, and recent evidence suggests that stress on natural resources from climate change can contribute directly and indirectly to migratory flows of people within and between states (Rigaud et al., 2018). Whatever the reason for movement, migration reduces the vulnerability of people to climate change. In the recent past, people have generally moved from more-vulnerable locations (i.e. from states that are highly exposed to climate impacts and which lack adaptation strategies) to less-vulnerable states (i.e. to states that have resources for climate change adaptation). As a result, international migration between 2010 and 2015 reduced human vulnerability to climate change, on average worldwide, by 15%

(Grecequet, DeWaard, Hellmann, & Abel, 2017). Yet, migration towards states with lesser vulnerability is not available to all, nor is it taking place in all regions of the world.

Human migration requires financial and social resources (wealth), and economic and social inequalities make migration impossible for those lacking these resources (Alvaredo, Chancel, Piketty, Saez, & Zuckman, 2018). Thus, climate vulnerability is negatively associated with social, political and economic capital (Black et al., 2011). The positively correlated relationship between migration and capital suggests that the ability to move is connected with availability of financial and social resources (McLeman & Smit, 2006). Applying these insights to the relationship between migration and climate vulnerability, Black et al. (2011) coined the term 'trapped populations' to refer to those who are unable to migrate from the most vulnerable places, in part, due to limited social and economic resources, and also those who are moving towards vulnerable places.

For example, despite political concern about migrants from Africa to Europe (World Report, 2017), most African migrants move within the continent (Abel & Sander, 2014; Laurence Flahaux & de Haas, 2016) and some are moving among states with similar or higher vulnerability (Grecequet et al., 2017). International migration of this type may not reduce the vulnerability of migrants. Studies conducted at the household level suggest that those who have enough income after a natural disaster, including disasters exacerbated by climate change, use that income for migration, but households that lack income and social resources post-disaster are prevented from migrating and are 'trapped' as well (Nawrotzki & DeWaard, 2018).

Though the term is used less often in natural systems, it is useful to think of trapped populations in nature as well. In natural systems, species need energy resources and appropriate cues to undertake migration. Dispersal ability naturally varies among taxa according to traits such as body size, lifespan and mode of movement. A meta-analysis of different species shifting their geographic ranges (the outcome of migration in one direction and population die-back in others) shows that many species are responding to a changing climate by moving towards the poles and higher elevation (Pecl et al., 2017). On average, terrestrial species have moved 17 km per decade, while marine species have moved 74 km per decade.

Where species come into contact with one another and hybridize, it is also possible to measure the rate of migration. A recent study suggests that two species of hybridizing butterfly, for example, have moved 40 km northwards in the last 30 years. If you compare observed, recent migration to what is likely to be needed under climate change during the twenty-first century, however, a considerable migration gap emerges. Davis and Shaw (2001), for example, estimate that climate change in the coming decades will necessitate range shifts of 300–500 km per century. This is a very long way for dispersal-limited species to travel, particularly when accounting for barriers to dispersal, such as habitat fragmentation, widespread agriculture and urbanization (Beever et al., 2016; Davis & Shaw, 2001; Schwartz et al., 2012).

Species that cannot keep up with this rate of change due to their own migration constraints or migration barriers are geographically trapped. Bumblebees provide a recent example of a species that has been trapped by climate change: their numbers are shrinking as they are especially sensitive to warming and lack capacity to adapt. For them, migration to the northern and cooler range is failing due to environmental factors such as differences in daylights and habitat changes including food availability and physiological factors such as a slow population growth rate (Kerr et al., 2015). We can observe that climate change is trapping bumblebee species in shrinking habitats and note that the loss of bee species has far-reaching consequences for ecosystems as well as people.

Thus, both, humans and other species can be trapped in unsustainable habitats. Such vulnerability is marked by strong patterns of social and economic inequality in human society and inequality in abundance in nature, such as the growth rate among bumblebee populations. Trapped populations and inequalities in their abilities to migrate are important policy concerns. In the following section, we discuss how vulnerability and adaptation are considered in different environmental and migration policy approaches for human and non-human populations.

4   Challenges of Migration Governance

Climate change is a chance for migration policy to evolve and modernize itself.
Dina Ionesco (2017)
International Migration Organization (IOM)

Climatic conditions during the early Holocene, when agriculture was first developed, were remarkably stable. Throughout much of human history, therefore, variation in climate has been small and, until recently, mean changes have been small and relatively slow. This climate stability has allowed for relatively static policy approaches to regulate the movements of people and species across political boundaries (Huntley, Hole, & Willis, 2011). Thus, biological conservation policies and laws tend to focus on traditional 'in-situ' or local conservation and retention of species in their historical ranges. Similarly, human migration policies have largely reflected nationalism and anti-migration sentiment (Postelnicescu, 2016), while planning for migration has largely focused on economic phenomena, such as migration from rural to urban areas (Massey, 1990). In many states, migration is perceived as a negative outcome rather than beneficial adaptation strategy (Melde, Laczko, & Gemenne, 2017; Nijenhuis & Leung, 2017). The pace of human-caused climate change, however, demands new policy approaches that acknowledge the need for human and non-human species to move beyond their traditional territories (Pecl et al., 2017). In Table 2 we chronologically summarize the key global and regional climate, environmental regimes, action plans and working groups that are addressing human and non-human species migration and that have emerged over the period of the past 70 years.

For almost half of the twentieth century, the global governance of human migration focused on forced movements driven by geopolitical factors (such as war and political prosecutions). The 1951 Refugee Convention was established as a response to the events in Europe after World War II (UNHCR, 2018), addressing 'forced migration of people' crossing a country border, and for the first time defined the term 'refugee' in its Article 1. A refugee is someone who is unable or unwilling

Table 2 Examples of global and regional frameworks, action plans and working groups addressing human migration and non-human species range shifts (migration); modified from Opitz Stapleton, Nadin, Watson and Kellett (2017). The list is not meant to be exhaustive

Human migration	Non-human migration
1951 Refugee Conventions	**1948** International Union for Conservation of Nature
1998 Guiding Principles on International Displacement	**1972** World Heritage Convention
2005 Hyogo Framework for Action	**1973** Convention on International Trade in endangered species (CITES)
2006 Global Migration Group	**1975** Ramsar Convention on Wetlands
2008 UN Convention to Combat Desertification (2008–2018 strategy)	**1979** UNEP Bonn Convention on Migratory Species and Wild animals
2010 UNFCCC Cancun Agreement	**1980** World Conservation Strategy
2012 Nansen Initiative	**1981** Japan-Australia Migratory Bird Agreement
2012 UN Advisory Group on Climate Change and Human mobility	**1990** South Pacific Apia Convention (suspended in 2006)
2015 Sendai Framework for Disaster Risk Reduction	**1992** European Union Habitat Directive and Birds Directive
2015 Addis Ababa Action Agenda	**1992** Convention on Biodiversity Protection
2015 Agenda for the Protection of Cross-border Displacement Persons in the Context of Disasters and Climate Change.	**1999** Trans-frontier Conservation Areas in Southern Africa (Southern Africa Development Community Protocol)
2015 2030 Agenda for Sustainable Development	**2005** Millennium Ecosystem Assessment
2015 UNFCCC Paris Agreement and Task Force on Displacement	**2010** Aichi biodiversity targets
2016 Platform on Disaster Displacement	**2012** Intergovernmental Science-Policy Platform on Biodiversity and Ecosystem services (IPBES)
2016 UN Summit for Refugees and Migrants	**2015** 2030 Agenda for Sustainable Development
2016 New Urban Agenda (UN-Habitat)	
2018 Global Compact for Migration	
2018 Global Compact for Refugees	

to return to their country of origin owing to a well-founded fear of being persecuted for reasons of race, religion, nationality, membership of a particular social group or political opinion. This rather narrow focus did not include climate factors that may force people—either voluntarily or involuntarily—to move away from home because such factors were negligible at the time (Piguet, 2013). Voluntary movement of people also falls under human rights laws, with migration law being largely within the responsibilities of national authorities. Currently, people moving because of either real or anticipated changes in their environments are not entitled to legal protections.

In natural systems, governments protect biodiversity largely through land set-asides and protected areas. Today, there are about 15% of terrestrial and inland waters, 10% of coastal and marine ecosystems and approximately 4% of global oceans designated as protected areas (UNEP-WCMC and IUCN, 2016). Early biodiversity conservation efforts of the International Union for Conservation for Nature (IUCN) were established in 1948 and focused primarily on designing protected areas and compiling a list of species at risk of extinction. Later different International Conventions were established with focus either on specific ecosystems (e.g. wetlands as reflected in the Ramsar Convention), sites (e.g. the World Heritage Convention) or type of species (e.g. the Convention on Migratory Species and wild animals).

The dependence of humans on biological diversity was considered for the first time within the Convention on Biological Diversity (CBD) in 1992. Later, trans-boundary protected areas and corridors were established across different states, including, for example, the Trans-Frontier Conservation Areas in Southern Africa established in 1999 as part of Southern Africa Development Community Protocol and the European Green Belt initiated in the early 1990s across twenty-four states. Still, protected areas are not well connected by bio-corridors, limiting the migration of species in response to changing conditions. Existing protected areas are too small and fragmented for maintaining viable populations. For example, Hallmann et al. (2017) found steady declines of insect biomass within protected areas in Germany.

A common feature of the human and non-human migration governance during the second half of the twentieth century was their limited focus on specific migrants: 'forced migrants due to geopolitical reasons' in human systems and on certain natural components, such as 'forest[s]' or groups of species, such as 'migratory birds' in non-human systems.

Much of the authority and responsibility have been in the hands of individual states, and the aims of governance in both systems were to protect and keep their populations largely in place. There was also little or no attention to climate-related migration prior to release of the first Intergovernmental Panel on Climate Change (IPCC) report in the early 1990s. The report clearly states that 'climate change could initiate large migration of people leading to severe disruption of settlements pattern and special instability' and that 'the climate zones which control species distribution will move pole-ward and to higher elevations'. In other words, the report suggested that both human and non-human species will face challenges due to climate change and that one possible strategy to mitigate the effects of climate change is migration. Following the IPCC, as shown in Table 2 increasing numbers of scientific or policy groups, task forces and action plans have emerged over the past two decades. In total, the efforts to address non-human migrations involve smaller numbers of activities. Interestingly, there are also differences in consideration of human and non-human migration within global climate and environmental regimes.

5   International Climate Change Regime

The United Nations Framework Convention on Climate Change (UNFCCC) is the major legal framework for international climate action and was adopted and signed during the Earth Summit in Rio de Janeiro in June 1992. During annual meetings of the Conference of Parties (COP), all states and parties to the Convention make decisions regarding implementation. The COP21 meeting in December 2015 led to the Paris Agreement in which each state submitted its plan to reduce greenhouse gas emissions and adapt to negative consequences of climate change. The concepts of vulnerability and adaptation have specific meanings and use in the Paris Agreement with respect to vulnerable populations and migrants. For instance: '*Parties* should, when taking action to address climate change, respect, promote and consider their respective obligations on human rights, the right to health, the rights of indigenous peoples, local communities, migrants, children, persons with disabilities and people in vulnerable situations and the right to development, as well as gender equality, empowerment of women and intergenerational equity'. It also makes reference to biodiversity protection and conservation: 'Noting the importance of ensuring the integrity of all ecosystems, including oceans, and the protection of biodiversity'. The Paris Agreement, however, does not specifically consider migration as an adaptation strategy for humans or non-human species. Lack of understanding and data on the relationships between climate and migration currently limits capacities to develop robust policy decisions and effective climate governance approaches.

To address this acknowledged but missing component, the UNFCCC established the Task Force on Displacement (TFD). This Task Force addresses the impact of extreme weather events on displacement and human migration and provides international negotiations on climate with recommendations on 'integrated approaches to avert, minimize and address displacement related to the adverse impact of climate change'. The Task Force was mandated to draw upon the work of, and involve, appropriate existing bodies and expert groups, including the Adaptation Committee, the Least-Developed Countries and other organizations outside of the UNFCCC, including the International Migration Organization (IMO), the International Labor Organization (ILO), the UN Development Program (UNDP), the UN High Commissioner for Refugees (UNHCR) and the Red Cross. This multi-agency effort aims to provide data to identify drivers of migration with a focus on climate change. Some perceive the establishment of the Task Force as a positive step forward, reflecting a need to involve multiple stakeholders to consider the multiple dimensions of human migration in decision-making processes regarding climate change. Others argue that the UNFCCC still lacks capacities and institutions to deal with human displacement and migration (Hodkinson & Young, 2012). One such area which lacks capacity is knowledge about the links between human migration and the impacts of species redistributions (range shifts) on food, fibre and fuel supplies.

Under the UNFCCC, national governments play a significant role in designing adaptation strategies. The Intended National Determined Contributions (INDC) statements submitted to the UNFCCC leading up to the Paris Agreement (Environmental Migration Portal, 2018) suggest that about 33 states refer to or consider climate-driven human migration and 7 states mention impacts of non-human migration of fish and marine mammals on their economic sectors (Environmental Migration Portal, 2018; UNFCCC, 2015). According to the Notre Dame Global Adaptation Index (NDGAIN), a tool that assesses climate vulnerability around the world (Chen et al., 2015), a majority of these states are among the most vulnerable to the impacts of climate change. In those most vulnerable circumstances, migration may be the only option.

Most INDCs focus on 'in situ adaptation' and disaster risk preparation and management rather than on promoting and managing migration of people within or between states. Many states also suggest rebuilding rural communities with the support of international funds, or remittances from diaspora migrants. The relocation of communities from high climate risk areas (flood zones) is also listed as an option. Interestingly, some states that refer to human migration also mention the effects of shifts in species ranges that are impacting upon food security and human health, such as Chad and Kiribati. Examples include changes in fish and mammal migration patterns that are already affecting fishery and tourism sectors.

6   The 2030 Agenda for Sustainable Development

Climate migration relationships are also relevant for attaining environmental and development goals, such as the 2030 Agenda for Sustainable Development adopted at the UN Sustainable Development Summit in 2015. This Agenda is a plan of action for people, planet and prosperity, which established 17 Sustainable Development Goals (SDG) and 169 targets, to be achieved by 2030 (United Nations, 2018). The Agenda and its goals build on earlier Millennium Development Goals (MDGs) and complete what they did not achieve. One of the omissions in the MDGs was the role of migration in achieving sustainable development. During the preparation of the 2030 Agenda, there was demand to include human migration and for the first time well-managed migrations were recognized as contributing to sustainable development as it is explicitly referenced in five of the seventeen Sustainable Development Goals. Migration is also a component of goals on gender equity, sustainable economic growth, peaceful and inclusive societies and data availability. One of the goals—Goal 10—is explicit and calls for 'reducing inequality within and among countries, including facilitation of orderly, safe, regular and responsible migration and mobility of people, implementation of planned and well-managed migration policies'. However, it does not give any specific quantitative targets that would suggest numbers of international migrants in the future (Abel, Barakat, Samir, & Lutz, 2016).

In contrast to references about human migration, the impact of species range shift on sustainable development is not accounted for in the SDGs (Pecl et al., 2017). Goals 14 and 15 (life below water and life on land) aim for biodiversity conservation and sustainable use of aquatic and terrestrial ecosystems, but they will only be successful if climate-driven change in species ranges and ecosystem consequences are taken into consideration. Pecl et al. (2017) also suggest that there are at least eleven goals that will be significantly affected by species movement. Interestingly, they correspond with four of the goals that explicitly consider human migration.

Both human and non-human migration will also affect Goal 13, which calls for climate action through climate resiliency and adaptive capacity building. Minimizing greenhouse gas emission and making climate change adaptation part of national action plans are essential components of this goal. However, human and non-human migration are not explicitly considered as an adaptation strategy. There is no doubt that human and non-human migration are interconnected and will affect one another, influencing whether and when climate targets and Sustainable Development Goals are achieved.

7   Blueprint of the New Migration Policy

Through a review of existing frameworks of migration policy, we have demonstrated that current policy approaches address the relationship between climate and migration in a limited way. Global refugee and biodiversity conservation laws provide protection only for specific groups of migrants (such as political refugees) and species (such as migratory species). Other forms of mobility and immobility caused by climate change (e.g. voluntary migration or trapped populations) are not fully addressed. International climate and sustainable development policies and regimes are starting to consider concepts of vulnerability and adaptation and the importance of migration for achieving their targets and goals. However, they focus on human migration only and conceive of migration as a problem to be managed and controlled rather than a process to be promoted (Nijenhuis & Leung, 2017). None of the frameworks legally acknowledge that migration can be an effective adaptation strategy for human and non-human species.

A new policy approach would address these limitations and move beyond traditional strategies that prevent or limit migration. The Sustainable Development Goals and the Paris Agreement, although imperfect, are important first steps in such a policy transition. We observe several distinct features about the relationships between climate change and migration to suggest a blueprint for new migration policies.

Many people will move voluntarily due to anticipated climate change. Such movement is currently in the hands of national authorities, and under current human rights and international law, these migrants are not entitled to special protections. However, part of the duty to receive migrants should include questions about climate vulnerability in their country of origin. Trans-boundary cooperation between states, similar to biodiversity protection such as the European Union's Habitats Directive and Trans-Frontier Conservation Areas in Southern Africa (Southern African Development Community Protocol, 1999), can help to guide voluntary movements of people through protecting the networks of human habitats across national borders. Ecologists are already discussing such novel conservation strategies and managed movement. Current national conservation policies and national plans are starting to address ways of helping species to navigate through highly fragmented landscapes by establishing short- and long-distance corridors that connect existing and sometimes rather isolated parks and reserves to receive migrating species, thereby allowing plants and animal species to move safely. Similarly, national governments could establish physical and/or socio-economic corridors for human migrants to safely reach their destinations and identify and enhance resilient communities that support diverse groups as destinations for climate migrants.

Trapped populations with limited mobility are not addressed in any of the instruments that govern or guide climate-related human mobility. Rather, the people who are trapped in the most vulnerable places and unable to adapt or move away from their homes become subjects of humanitarian aid. In contrast, biodiversity conservation law acknowledges that a large number of species are susceptible to extinction due to climate change if they are unable to adapt or shift their distribution (e.g. in the IUCN red list of Threatened Species and the US Endangered Species Act (ESA)). These species often become priorities of conservation policies and the focus of adaptation planning, where migration is formally considered as an option (Lopez, 2015). Human migration governance could similarly identify ways to safeguard trapped populations and identify hotspots where populations can become trapped. In doing this work, ecologists, social scientists and policy makers must balance the potential risks and benefits of managed relocations. In some cases, relocation outside a known historical distribution is the only option for survival, but doing so will have risks and side effects. Relocations in the absence of climate change, for example, have resulted in severe consequences historically, such as local extinctions of native species (non-human) and increased social and economic vulnerabilities for human migrants (Melde et al., 2017). Relocation decisions will be difficult to reach because of current limitations to predict both vulnerability and the consequences of relocations. However, managed relocations have already been conducted for certain endangered species, such as in responding to harmful impacts of invading species. Examples include the Torreya Guardians, a conifer tree, that began to decline due to fungal pathogens (torreyagardians.org) and island communities, such as the Vunidogoloa village in Fiji, which moved two kilometres inland to shift away from eroding coastline (Tronquet, 2014). These examples can be useful starting points and experiences in developing decision frameworks and guidelines.

The governance of migration of both human and non-human species requires collaboration between legally binding laws under the

UN umbrella (such as the Paris Agreement, SDGs, Convention on Biodiversity Conservation) and non-legally binding instruments, such as non-treaty agreements, declaration, guidance, principles or conclusion of Executive Committees. Non-binding instruments are easier to negotiate and include various stakeholders that do not participate during international negotiations (Ferris & Bergmann, 2017). An example of non-binding efforts in managing human migration is the Nansen Initiative, a state-led process based on participatory consultation that developed an agenda for the protection of cross-border displaced populations due to climate disasters. There are other types of non-binding instruments helping national governments to address different forms of migration including the Guiding Principles and Internal Displacement of the UN Commission on Human Rights. In the management of non-human migration, a range of non-legally binding instruments have been developed to address translocation of species, including human-mediated movements of species from one area to another to establish new populations or restore key ecological functions (e.g., see, IUCN/SSC, 2013). Yet, there are far fewer non-binding instruments in biodiversity conservation in comparison with human migration policy (Kotze, 2014).

A policy approach based on a combination of legally and non-legally binding instruments that involves the views and perspectives of different actors (such as policy makers, NGOs and publics) dilutes centralized responsibilities and authority, potentially making it difficult to hold different actors accountable for their decisions and actions (Mees & Driessen, 2018). Yet, such an approach can provide tools, including guidelines based on vulnerability assessments, to hold governments accountable and allow for broad participation in diffusing processes that might benefit from being experimental and uncoordinated, at least initially. Accountability with respect to governance of climate change adaptation has not yet been studied (Mees & Driessen, 2018), and more research is needed to identify and evaluate the efficacy of different approaches, including approaches for managing human and non-human migration.

The threats of climate change for human societies and biodiversity are not likely to be solved by traditional 'in situ adaptation' approaches, and small adjustments or changes to current policy will not provide long-term answers. We can expect that more people, animal and plant species will be on the move or trapped by climate change in the near future. Many will be at risk in their cross-border movements if no actions are undertaken. We hope that thinking of migration broadly—across both human and non-human species—helps to inspire creative solutions and new approaches for the benefit of all.

References

Abel, G. J., & Sander, N. (2014). Quantifying global international migration flows. *Science, 343*(6178), 1520–1522.

Abel, G. J., Barakat, B., Samir, K. C., & Lutz, W. (2016). Meeting the sustainable development goals to lower world population growth. *Proceedings of National Academy of Sciences, 113*(50), 14294–14299. https://doi.org/10.1073/pnas.1611386113.

Alverado, F. M., Chancel, L., Piketty, T., Saez, E., & Zuckman, G. (2018). *World inequality report*. World Inequality Lab. http://wir2018.wid.world/files/download/wir2018-full-report-english.pdf. Accessed May 2, 2018.

Beever, E. A., O'Leary, J., Mengelt, C., West, J. M., Julius, S., Green, N., …, Hofmann, G. E. (2016). Improving conservation outcomes with a new paradigm for understanding species fundamental and realized adaptive capacity. *Conservation Letter, 9*(2), 131–137.

Black, R., Bennett, S. R., Thomas, S. M., & Beddington, J. R. (2011). Climate change: Migration as adaptation. *Nature, 478*, 447–449.

Blows, M. W., & Hoffmann, A. A. (2005). A reassessment of genetic limits to evolutionary change. *Ecology, Ecological Society of America*. https://doi.org/10.1890/04-1209.

Cardona, O. D., van Aalst, M. K., Birkmann, J., Fordham, M., McGregor, G., Perez, R., …, Sinh, B. T. (2012). Determinants of risk: Exposure and vulnerability. In C. B. Field, V. Barros, T. F. Stocker, D. Qin, D. J. Dokken, K. L. Ebi, …, P. M. Midgley (Eds.), *Managing the risks of extreme events and disasters to Advance climate change adaptation. A special report of working groups I and II of the Intergovernmental Panel on Climate Change (IPCC)* (pp. 65–108). Cambridge and New York, NY: Cambridge University Press.

Chen, C. Noble, I., Hellmann, J., Coffee, J., Murillo, M., & Chawla, N. (2015). *Univerity of Notre Dame Global Adaptation Index, Country Index*

Technical Report. Available at https://gain.nd.edu/assets/254377/nd_gain_technical_document_2015.pdf.

Clark, J. S., Fastie, C., Hurtt, G., Jackson, S. T. Johnson, C., King, G. A., ..., Wycoff, P. (1998). Reid's paradox of rapid plant migration. Dispersal theory and interpretation of paleoecological records. *Bioscience, 48*, 13–24.

Convention on Biological Diversity (CBD). (2017). *Review of future projections of biodiversity and ecosystem services.* Subsidiary Body on Scientific and Technological Advice, 21 Meeting, Montreal, December 11–14. Available at https://www.cbd.int/sbstta/sbstta-22-sbi-2/sbstta-21-inf-02-en.pdf.

Cresswell, K. A., Satterhwaite, W. H., & Sword, G. A. (2011). Understanding the evolution of migration through empirical examples. In E. J. Milner-Gulland, J. M. Fryxell, & A. R. E. Sinclair (Eds.), *Animal migration, a synthesis.* Oxford: Oxford University Press.

Darwin, C. (1859). *On the origin of species by means of natural selection.* London, UK: John Murray Albemare st.

Davis, M. B., & Shaw, R. G. (2001). Range shifts and adapative responses to quaternary climate change. *Science, 292*(5517), 673–679.

De Vos, J. M., Joppa, L. M., Gittleman, J. L., Stephen, P. R., & Pimm, S. L. (2015). Estimating the normal background rate of species extinction. *Conservation Biology, 29*(2), 452–462. https://doi.org/10.1111/cobi.12380,2015.

Deutsch, C. A., Tewksbury, J. J., Huey, R. B., Sheldon, K. S., Ghalambor, C. K., Haak, D. C., & Martin, P. R. (2008). Impact of climate warming on terrestrial ectotherms across latitude. *PNAS, 105*(18), 6668–6672.

Environmental Migration Portal. (2018).http://www.environmentalmigration.iom.int/migration-indcsndcs.

Ferris, E., & Bergmann, J. (2017, March). Soft law, migration, and climate change governance. *Journal of Human Rights and the Environment, 8*(1), 6–29.

Geisler, C., & Currens, B. (2017). Impediments to inland resettlements under conditions of accelerated sea level rise. *Land Use Policy, 66*, 322–330.

Grecequet, M., DeWaard, J., Hellmann, J. J., & Abel, G. J. (2017). Climate vulnerability and human migration in global perspective. *Sustainability, 9*, 720.

Hallmann, C. A., Sorg, M., Jongejans, E., Siepel, H., Hofland, N., Schwan, H., ..., de Kroon, H. (2017). More than 75 percent decline over 27 years in total flying insect biomass in protected areas (E. G. Lamb, Ed.). *PLoS One, 12*(10), e0185809. https://doi.org/10.1371/journal.pone.0185809.

Hellmann, J. J., & Pineda-Krch, M. (2007). Constraints and reinforcement on adaptation under climate change: Selection of genetically correlated traits. *Biological Conservation, 137*(4), 599–609.

Higgins, S. I., & Richardson, D. (1999). Predicting plant migration rates in changing world: The role of long-distance dispersal. *American Naturalist, 153*(5), 464–475.

Hill, J. K., Griffits, H. M., & Thomas, C. D. (2011). Climate change and evolutionary adaptations at species range margins. *Annual Review of Entomology, 56*, 143–159.

Hodgkinson, D., & Young, L. (2012). *In face of looming catastrophe: A convention for climate change displaced persons.* Available at http://ccdpconvention.com.

Hoffman, A. A., & Sgro, C. M. (2011). Climate change and evolutionary adaptation. *Nature, 470*, 479–485.

Hunter, L., & Nawrotzki, R. (2016). Migration and environment. In M. J. White (Ed.), *International handbook of migration and population distribution.* Springer.

Huntley, B., Hole, D. G., & Willis, D. J. (2011). Assessing the effectiveness of a protected area network in the face of climate change. In T. R. Hodkinson, M. B. Jones, S. Waldren, & J. A. N. Parnell (Eds.), *Climate change, ecology, and systematics.* Cambridge: Cambridge University Press.

Husby, A., Visser, M. E., & Kruuk, L. E. B. (2011). Speeding up microevolution: The effects of increasing temperature on selection and genetic variance in wild bird population. *PLoS Biology, 9*(2), e1000585. https://doi.org/10.1371/journal.pbio.1000585.

Ionesco, D. (2017). http://www.transre.org/en/blog/climate-change-migration-policy-2017-bonn-intersessional/.

IPCC. (2012). *Managing the risks of extreme events and disasters to advance climate change adaptation; A special report of working groups I and II of the intergovernmental panel on climate change* (p. 582). Cambridge, UK; New York, NY, USA: Cambridge University Press.

IPCC. (2014). Summary for policymakers. In C. B. Field, V. R. Barros, D. J. Dokken, K. J. Mach, M. D. Mastrandrea, T. E. Bilir, …, L. L. White (Eds.), *Climate change 2014: Impacts, adaptation, and vulnerability. Part A: Global and sectoral aspects. Contribution of working group II to the Fifth Assessment Report of the Intergovernmental Panel on Climate Change* (pp. 1–32). Cambridge, UK and New York, NY, USA: Cambridge University Press.

IUCN/SSC. (2013). *Guidelines for reintroductions and other conservation translocations. Version 1.0* (viiii + 57pp). Gland: IUCN Species Survival Commission.

Javeline, D., Hellmann, J. J., McLachlan, J. S., Sax, D. F., Schwartz, M.W., & Cornejo, R. C. (2015). Expert opinion on extinction risk and climate change adaptation for biodiversity. *Elementa Science of the Anthropocene, 3,* 57. http://doi.org/10.12952/journal.elementa.000057.

Kerr, J. T., Pindar, A., Galpern, P., Packer, L., Potts, S. G., Roberts, S. M., …, Pantoja, A. (2015). Climate change impact on bumblebees converge across continents. *Science, 349*(6244), 177–180. https://doi.org/10.1126/science.aaa7031.

Kotze, L. J. (2014). Transboundary environmental governance of biodiversity in anthropocene. In L. J. Kotze & T. Marauhn (Eds.), *Transboundary governance of biodiversity.* ProQuest Ebook Central. Accessed May 2018.

Laurence Flahaux, M., & de Haas, H. (2016). African migration: Trends, patterns, drivers. *Comparative Migration Studies, 4*(1).

Lee, E. (1966). A theory of migration. *Demography, 3*(1), 47–57.

Lenoir, J., & Svenning, J. C. (2014). Climate-related range shifts-global multi-dimensional synthesis and new research directions. *Ecography, 38*(1), 15–28.

Locke, J. T. (2009). Climate change induced migration in the Pacific region: Sudden crisis and long-term developments. *The Geographical Journal, 178*(3), 171–180.

Lopez, J. (2015). Biodiversity on the brink: The role of assisted migration in managing endangered species threatened with rising seas. *Harvard Environmental Law Review, 39*(1). Available at http://harvardelr.com/print-archives/.

Massey, D. (1990). The social and economic origins of immigration. *The Annals of American Academy of Political and Social Science, 510*(1), 60–72.

McLeman, R. (2014). *Climate and human migration. Past experience, future challenges.* New York: Cambridge University Press.

McLeman, R., & Smit, B. (2006). Migration as adaptation to climate change. *Climatic Change, 76*(1–2), 31–53.

Mees, H., & Driessen, P. (2018). A framework for assessing the accountability of local governance arrangements for adaptation to climate change. *Journal of Environmental Planning and Management.* https://doi.org/10.1080/09640568.2018.1428184.

Melde, S., Laczko, F., & Gemenne, F. (2017). *Making mobility work for adaptation to environmental change. Results fro the MCLEP global research.* Le Grand-Saconnex: International Organization for Migration.

Nathan, R. (2006). Long-distance dispersals of plants. *Science, 313,* 786–788.

Nawrotzki, R., & DeWaard, J. (2018). Putting trapped population into place: Climate change and itenr-district migration flow in Zambia. *Regional Environmental Change.* https://doi.org/10.1007/s10113-017-1224-3.

Nijenhuis, G., & Leung, M. (2017). Rethinking migration in the 2030 agenda: Towards a de-territorialized conceptualization of development. *Forum for Development Studies, 44*(1), 51–68.

Opitz Stapleton, S., Nadin, R., Watson, C., & Kellett, J. (2017). Climate change, migration, and displacement. The need for a risk-informed and coherent research. ODI/UNDP report, November 2017.

Pecl, G. et al. (2017). Biodiversity redistribution under climate change: Impact on ecosystems and human well-beings. *Science, 355.* https://doi.org/10.1126/science.aai9214.

Piguet, E. (2013). From primitive migration to climate refugees. The curious fate of natural environment in migration studies. *Nature and Society, 103*, 148–162.

Postelnicescu, C. (2016). Europe's new identity: The refugee crisis and rise of nationalism. *European Journal of Psychology, 12*(2), 203–209. https://doi.org/10.5964/ejop.v12i2.1191.

Ratzel, F. (1882). *Antropogeographie.* Stutgart.

Ravenstein, E. G. (1885). The laws of migration. *Journal of the Statistical Society of London, 48*(2), 167–235.

Richardson, D. M., Hellmann, J. J., McLachlan, J. S., Sax, D. F., Schwartz, M. W., Gonzalez, P., …, Vellend, M. (2009). Multi-dimensional evaluation of managed relocation. *Proceedings of the National Academy of Sciences of the United States of America, 106*(24), 9721–9724.

Rigaud, K. K., de Sherbinin, A., Jones, B., Bergmann, J., Clement, V., Ober, K., …, Midgley, A. (2018). *Groundswell: Preparing for internal climate migration.* Washington, DC: World Bank.

Rowlands, A. (2018). *Response to 'A new era of human and non-human migration. Radical ecological conversion after Laudato si. Discovering the intrinsic value of all creatures, human and non-human'.* Rome, March 7–8.

Ryan, S. F., Deines, J. M., Scriber, J. M., Pfrender, M. E., Jones, S. E., Emrich, S. J., & Hellmann, J. J. (2018). Climate-mediated hybrid zone movement revealed with genomics, museum collection, and simulation modeling. *Proceedings of the National Academy of Sciences,* 20171495. https://doi.org/10.1073/pnas.1714950115.

Samberg, L. H., Gerber, J. S., Ramakutty, N., Herrero, M., & West, P. (2016). Sub-national distribution of average farm-size and smallholder contributions to global food production. *Environmental Letters, 11*(12), 124010.

Schwartz, J. (2005). Who is sensitive to extremes of temperatures? A case-only analysis. *Epidemiology, 16*, 67–72.

Schwartz, M. W., Hellmann, J. J., McLachlan, J. S. (2009). The precautionary principle in managed relocation is misguided advice. *Trends in Ecology and Evolution, 25*, 474.

Schwartz, M. W., Hellmann, J. J., McLachlan, J., SAX, D. F., Borevitz, J. O. Brennan, J., ..., Zellmer S. (2012). Managed relocation: Integrating the scientific, regulatory and ethical challenges. *Bioscience, 62*(8), 732–743.

Stafoggia, M., Forastiere, F., Agostini, D., Biggeri, A., Bisanti, L., Cadum, E., ..., Perucci, C. A. (2006). Vulnerability to heat related mortaility. *Epidemiology, 17*(3).

Tronquet, C. (2014). From Vunidogoloa to Kenani: An insight into successful relocation. In F. Gemenne, C. Zickgraf, & D. Ionesco (Eds.), *The state of environmental migration 2015. A review of 2015.* IMISCOE.

UNEP-WCMC and IUCN. (2016). *Protected planet report 2016.* Cambridge and Gland: UNEP-WCMC and IUCN.

UNFCCC. (2015). *Paris agreement.* Available at https://unfccc.int/sites/default/files/english_paris_agreement.pdf.

UNHCR. (2018). United Nation Refugee Agency. *Convention and protocol relating to the status of refugees.* Available at http://www.unhcr.org/en-us/3b-66c2aa10. Accessed May 2018.

United Nations. (2018). *Sustainable development knowledge platform.* Available at https://sustainabledevelopment.un.org/sdgs. Acessed June 18, 2018.

Wagner, M. (1873). *The Darwinian Theory and the law of the migration of organisms.* London: Edward Stanford, 6 and 7, Charing. Cross, S.W.

White, M. J. (2016). *International handbook of migration and population distribution.* Springer, ISBN 978-94-017-7281-5. https://doi.org/10.1007/978-94-017-7282-2.

Williams, S. E., Shoo, L. P., Isaac, J. L., Hoffmann, A. A., & Langham, G. (2008). Towards an integrated framework for assessing the vulnerability of species to climate change. *PLoS Biology, 6*(12), e325. https://doi.org/10.1371/journal.pbio.0060325.

Wilson, R. P., Hustler, K., Ryan, P. G., Burger, A. E., & Noldeke E. C. (1992). Diving birds in cold water: Do Archimedes and Boyle determines energy costs? *The American Naturalist, 140*(2).

World Report. (2017). *Events of 2016, human rights watch.* https://www.hrw.org/sites/default/files/world_report_download/wr2017-web.pdf.

Zelinsky, W. (1971). The hypothesis of the mobility transition. *Geographical Review, 61*(2), 219–249.

10

Representing Whose Access and Allocation Interests? Stakeholder Perceptions and Interests Representation in Climate Governance

Timothy Cadman, Tek Maraseni,
Hugh Breakey and Hwan-ok Ma

1   Introduction

Given the distance from power for some stakeholders, ensuring that relevant interests are represented in global venues is a considerable governance challenge (Stiglitz, 2003, p. 61). Stakeholder inclusion, or 'inclusiveness', is a critical aspect of interest representation at the global level

T. Cadman · H. Breakey (✉)
Griffith University, Nathan, QLD, Australia
e-mail: h.breakey@griffith.edu.au

T. Cadman
e-mail: t.cadman@griffith.edu.au

T. Maraseni
University of Southern Queensland, Toowoomba, QLD, Australia
e-mail: Tek.Maraseni@usq.edu.au

H. Ma
International Tropical Timber Organization, Yokohama, Japan
e-mail: ma@itto.int

© The Author(s) 2019
B. Edmondson and S. Levy (eds.), *Transformative Climates and Accountable Governance*, Palgrave Studies in Environmental Transformation, Transition and Accountability, https://doi.org/10.1007/978-3-319-97400-2_10

223

and varies greatly depending on the institutional context in question (Koenig-Archibugi, 2006, p. 13). There has been recognition for some time that globalisation, and the post-modern experiences of complex governance it has brought about, requires a reconfiguration of interest representation in contemporary public administration (Rhodes, 1997, p. 198). In the supranational polity of the EU, for example, inclusiveness has been identified as a fundamental principle of 'good' governance and legitimacy (Schmidt, 2013, pp. 2–3; Smismans, 2004, p. 26).

Governance scholars have broken down inclusiveness into two concepts: weight and access. 'Weight', which may be linked to the notion of equality, implies the degree of influence that participants are able to exert in a given policy context: the more balanced the power, the greater the equality between participants. 'Access' denotes the range of stakeholders and the amount of influence they may exert in policy content-development. Governance systems can therefore be determined to be inclusive when stakeholders are not only involved in decision-making, but their input is both formally and informally acknowledged in actual policy (Koenig-Archibugi, 2006, pp. 14–15). It is argued that interest representation is most inclusive when previously excluded or marginalised voices and perspectives are actively encouraged (Young, 2000, p. 8). This work on governance dovetails with the recent 'procedural turn' in the literature on climate justice. This new area of ethical focus eschews commentary on the equitable distribution of mitigation/adaptation costs and benefits, instead exploring fair terms for international discourse and the procedural justice required by 'public reason', whereby nations develop through norms, principles and policies through fair terms of deliberation (Boran, 2016; Boran & Katz, 2017; Breakey, 2015).

Seeking the involvement of such interests has little meaning if the resources for their participation are not available. Traditionally, it is only the most well-resourced organisations that have the capacity to attend and be involved in global policy deliberations. This limitation may be mitigated if there are compensating opportunities from groups to collaborate. In this regard, the growth of network-models of governance has been seen as providing a degree of compensation for those groups that cannot attend global (international) venues of power—such as the UNFCCC Conference of Parties (COP). This is, of course, contingent on the resourcing of those networks through which the absent interests are represented (Scholte, 2004, pp. 223–225).

Scholars identify a wide range of resources required for effective interest representation. Clearly, economic resources are required (for such basic necessities as travel, food and accommodation). Technical capacity (including know-how, experience and expertise) is also important, as is institutional support (Mason, 1999, pp. 72–73). These have all been identified as structural framework conditions for the development of good (effective) environmental policy for quite some time (Jänicke, 1992, 1996). Beyond the need for economic, technical and institutional capacity building, there is also a need for the appropriate distribution of the monetary and non-monetary benefits created by the implementation of those policies in order for governance systems to function effectively. These factors are pronounced during policy formulation, such as has been observed in the policy arena of emissions reduction activities such as REDD+. Finally, benefit sharing throws up its own governance challenges, particularly in terms of who receives those benefits (Chapman, Wilder, & Millar, 2014, p. 271).

While international climate initiatives can be seen to be inclusive of various stakeholder interests, the allocation of resources to facilitate their active participation continues to be problematic. Previous studies suggesting the existence of a 'South/North' divide in global climate governance appear to be confirmed, although this observation is qualified. These findings have implications for the future implementation of activities aimed at reducing global greenhouse gas emissions under the 2015 Paris Agreement, and for any new market mechanisms that eventuate, notably the so-called sustainable development mechanism (SDM).

This chapter presents a synthesis of findings from quantitative and qualitative investigations of the perspectives of participants involved in international climate governance, conducted over the period 2010–2015. In this study, an established framework of principles, criteria and indicators (PC&I) for institutional governance was applied to two mechanisms under the United Nations Framework Convention on Climate Change (UNFCCC): the initiative referred to as 'Reducing emissions from deforestation and forest degradation and the role of conservation, sustainable management of forests, and enhancement of forest stocks in developing countries' (REDD+) and the Clean Development Mechanism (CDM) of the Kyoto Protocol (KP).

Assessment focuses on the governance value of interest representation in terms of inclusiveness (access) and resources (allocation). It begins by outlining the historical context of UNFCCC, as well as CDM and REDD+, and continues with a delineation of the methods adopted and results to reveal a relatively consistent set of results across the elements investigated, with inclusiveness receiving the highest score of all the governance indicators and resources the lowest. The CDM was the weakest performer.

2   Challenges Confronting Climate Governance

A range of academics and commentators has pointed to governance problems with the programmes and mechanisms of international climate governance (Cadman, 2013; Forsyth, 2009; Hoffmann, 2011; Knieling & Leal Filho, 2012; Lederer, 2011; Lyster, 2011; Thompson, Baruah, & Carr, 2011). Both CDM and REDD+ have been identified as sites of political negotiation and contestation, wherein politically charged concepts such as environmental integrity and equity become turned into technical matters, referred to as 'governing by expertise' (Dooley & Gupta, 2017). The role of governance, notably in the CDM, has been linked to the political-economic context in which projects are operating; this in turn heavily influences the quality of the outcomes generated, particularly in developing states, where there is a need to broaden the capacity of projects to improve their effectiveness (Newell, 2009). Governance has been the focus of several hundred academic papers on REDD+ (Cadman, Maraseni, Breakey, López-Casero, & Ma, 2016), particularly regarding the vexed issue of how to share the benefits derived from reducing emissions (Chapman et al., 2014; Chapman, Maguire et al. 2015; Chapman, Wilder et al., 2015; Harada, Prabowo, Aliadi, Ichihara, & Ma, 2015; McGregor et al., 2014). According to commentators and scholars, the so-called North/South divide between developed and developing states has been exacerbated in international climate governance due to arguments around the use of forest carbon under CDM and REDD+ (Abreu Mejía, 2010; Allan & Dauvergne, 2013; Bäckstrand & Lövbrand, 2007; Roberts & Parks, 2006). Given the significant scale of finance for REDD 'readiness', and the potential for results-based payments to generate similar levels of funding, it may be better to call this a 'South North divide', as significant resources are flowing from developed to developing states, with concomitant impact on perceptions of both REDD+ and CDM (Cadman, 2013).

The investigation below evaluates CDM and REDD+ against a comprehensive set of governance values, but concentrates largely on the nature of interest representation, and the related concepts of access (understood in terms of the inclusiveness of governance structures) and allocation (understood in terms of the provision of resources, capacity building and benefit sharing). It looks at the views of a range of participants, but focuses especially on evaluating these perceptions from developed and developing country perspectives.

3   Key Developments and Policies: Overview

3.1   Background: UNFCCC and its Institutional Arrangements for Climate Governance

The United Nations Framework Convention on Climate Change (UNFCCC) entered into force in 1994 with the aim of preventing 'dangerous anthropogenic interference with the climate system' (Article 2) and is one of the most concrete outcomes of the 1992 UN Conference on Environment and Development (UNCED). The Convention was initially operationalised via the 1997 Kyoto Protocol (KP) in a time when the market- and political ideology of neoliberalism was in its ascendency. The Kyoto Protocol's market emphasis was demonstrated by its three main policy instruments (or 'flexible mechanisms') aimed at mitigating emissions of greenhouse gases (GHG) via the Clean Development Mechanism (CDM), allowing developed states to 'offset' their emissions via projects in developing states, Joint Implementation (JI), a similar model aimed at economies in transition to a market economy, and International Emissions Trading (IET), which was an attempt to link all these activities within an international carbon offsetting economy. The European Union was the first collective polity to enter into the system in 2005 through its own Emissions Trading Scheme (EU ETS), which from 2008 permitted the purchase of 'carbon credits' (referred to as certified emissions reductions or CERs) from projects established under the CDM, as well as JI projects (Cadman, 2013; Cadman et al., 2015).

The significance of the Kyoto Protocol was that it stipulated a target that was legally binding for the reduction of greenhouse gases (most notably carbon dioxide) of a minimum of 5.2% over the timeframe of 2008–2012, using 1990 as a reference level. The CDM is the only one of the three mechanisms that can be said to be genuinely global in its institutional expression (Maraseni & Cadman, 2015). This is because it transfers both technology and direct project finance from developed states (referred to as 'Annex I') to developing (or 'Non-annex') states. The underlying principles were that developed states could enable sustainable development (an objective of UNCED) by facilitating the exchange (offsetting) of emissions from private sector industry via projects in developing states (thereby meeting their own national greenhouse gas reduction targets) and removing atmospheric pollution that would not have happened without such initiatives (a Kyoto Protocol requirement referred to as 'additionality'). The first CDM project was registered in 2004 in Brazil, with the number of activities registered growing from 62 to over 7,000 between 2005 and the end of the commitment period, offsetting over one billion tonnes of carbon. Reducing or avoiding emitting activities (referred to as 'mitigation') was initially the primary focus of activities under the Convention, but this changed over the period 2001–2006, between Conference of Parties (COP) 7 (Marrakesh) and COP 12 (Nairobi), as developing states began to focus on their exposure (or vulnerability) to climate change, and the costs that coping with (referred to as 'adaptation') climate change would entail.

As a result, climate change began to be reframed as not just an environmental issue, but a social one as well, which needed the provision of resources, usually financial, and often somewhat euphemistically referred to as 'capacity building'. While this change in emphasis did not particularly affect how states began implementing the Kyoto Protocol, it introduced a lack of certainty into the negotiations about what the policy settings would look like after 2012, and the extent to which market mechanisms would continue to be used for mitigation activities. Beyond what might be interpreted as an attempt to merely gain extra funds from developed states was the inherent problem that the Kyoto Protocol only focused on developed states' obligations, while several states—most notably China and India, the largest beneficiaries of CDM projects to the virtual exclusion of all others—changed their developmental status over the period, yet remained outside any of the protocol's emissions reductions commitments (Cadman et al., 2015). In other words, the CDM's aim was to serve as a mechanism for aiding sustainable development through financial transfer from high income and high emitting states to low income and low emitting states. However, as India and (especially) China moved out of the latter category, the mechanism became increasingly untenable.

This underlying tension came to boil in 2010 at COP 15 (Copenhagen), where there were overly optimistic expectations for a legally binding new arrangement, but which ultimately failed to deliver anything except a commitment for the Kyoto Protocol to continue in some form, and the vague idea that all states might reduce emissions through what were referred to as 'intended nationally determined contributions' (INDCs) (Radunsky, 2017). The policy uncertainties generated prior to the expiration of the Kyoto Protocol commitment period resulted in the CDM being declared to be 'imperilled' (CDM-Policy-Dialogue, 2012a, p. (a)2), with further predictions at the time that 'if nations permit the CDM mechanism to disintegrate, the political consensus for truly global carbon markets may evaporate' (CDM-Policy-Dialogue, 2012b, p. (b)). Nevertheless, at COP 18 in 2012 the 'Doha' amendment (DA) to the Kyoto Protocol resulted in agreement for a second commitment period from 2013 to 2020, during which the (by now) 38 signatories would implement emissions reductions of 18%. In terms of total emissions globally, however, it should be stressed that the Doha Amendment addresses a mere 15% of emissions, and a number of the original signatories to the protocol did not sign up (Maguire, 2015). With the recognition implicit in the Doha Amendment that erstwhile developing states also had obligations to reduce emissions, new meaning was given to the UN norm of 'common but differentiated responsibilities' (CBDR) between developing and developed states, and the notion of a 'new market mechanism' began to gain traction, which might redefine the CDM to include developing states, while continuing to serve developed states' interests. However, there were two hurdles to overcome. The first was in some regions there was an ideological rejection of neoliberal market mechanisms (especially in Latin America). The second was the growing attraction of financing emissions reductions from improved forest management activities, which had been largely excluded from the CDM, which focused on planted forests, rather than dealing with deforestation and forest degradation for logging activities in natural forests (Cadman et al., 2015).

This was a deliberate omission on behalf of the CDM, which had been under pressure from NGOs in the early days of the Kyoto Protocol negotiations not to include forests. In 2005, the idea of 'reducing emissions from deforestation' (RED) entered the climate negotiations in the form of an emerging policy discussion. This effectively reinserted forests into the mitigation negotiations, and the proposed mechanism and related programme evolved over time to include 'reducing emissions from deforestation *and forest degradation*' (REDD), to REDD-'plus' at COP 15, to include other non-forestry activities. The idea behind the initiative was to combat forest loss by providing funds to forest actors (largely in the developing and tropical rainforest states) to reduce their impacts by managing forests more sustainably, either through conservation measures, maintaining or enhancing stocks of carbon, or reducing logging impacts. This would provide funds for developing states and at the same time allow developed states to continue offsetting their own emissions through some form of carbon trade, but discussions as to how this might occur have constantly changed over time (de Oliveira et al., 2013). Rather than delivering tangible results on the ground, payments that have been made to date have been focused largely on assisting developing states 'in building their capacity to benefit from possible future systems of positive incentives for REDD+' (FCPF, 2017).

The initiative, referred to in climate negotiations under the long title of 'Reducing emissions from deforestation and forest degradation and the role of conservation, sustainable management of forests and enhancement of forest carbon stocks in developing countries', is supported by a range of UN-affiliated entities. A support agency, UN-REDD, provides help to identified states during the introduction of the programme, referred to as 'REDD+ readiness', through a number of regional and global funds allocated by donor states. UN-REDD operates in collaboration with the UN Development Programme (UNDP), the UN Environment Programme (UNEP) and the Food and Agriculture Organisation (FAO) (UN-REDD, 2009). The UN also operates a Multi-Partner Trust Fund Office (MTFO) which allocates and tracks the use of developed state donations, which are distributed through a number of specific programmatic funds, which between 2008 and 2017 totalled more than USD \$3.5 billion. EU states provided a significant proportion of money, as well as Canada, Japan and Switzerland. Norway is one of the largest individual donor states.[1] The World Bank also plays a role, via its Forest Carbon Partnership Facility (FCPF), and manages two funds—the Readiness Fund, which complements UN-REDD support, and the Carbon Fund, which develops funding arrangements for the future possible purchase and trade of emissions reductions. As an institution, FCPF is made up of state donors, NGOs and the private sector and concentrates on the more technical requirements or REDD+, including Readiness Preparation Proposals (RPPs) and Emissions Reduction Programs (ERPs). Pledges and allocated funds have exceeded \$1 billion so far (FCPF, 2017).

A range of further agreements has been negotiated between developed and developing states under the auspices of the climate negotiations. In 2009, the Copenhagen Accord of COP 15 stipulated that a tranche of funds from the \$100 billion allocated to climate change management should go to emissions reduction activities in forests (Maraseni & Cadman, 2015). Donating and accounting for such potentially large funds pose a number of governance challenges for the

---

[1]See http://mptf.undp.org/tools/search?q=REDD&qc=project (accessed 8/9/2017).

donor and recipient states, as recognised in Cancún in 2010, at COP 16. Parties to the Convention agreed to the creation of 'guidance and safeguards' to ensure 'transparent and effective national forest governance', including the 'full and effective participation of relevant stakeholders, in particular indigenous peoples and local communities', with the usual caveat that this was conditional on 'national legislation and sovereignty' (UNFCCC, 2011, p. 26). The negotiations further recognise that 'land tenure issues, forest governance issues, [and] gender' need to be taken into consideration (Article 72). Additional efforts to safeguard the needs of stakeholders were announced by the REDD+ related agencies themselves in 2011 at COP 17 in Durban, and a number of guidelines were released (FCPF, 2011; UN-REDD, 2012). Together, these governance provisions were further systematised at COP 19, 2013, via the Warsaw Framework for REDD+ (UNFCCC, 2014). The otherwise unstoppable policy momentum of the initiative was briefly checked in Lima 2014 during COP 20, which may be attributed to Parties hedging their bets in anticipation of the outcomes of the forthcoming Paris COP (2015), where the fate of the CDM, and future market mechanisms, was expected to be determined (Radunsky, 2017).

The top-down model of the Kyoto Protocol was considerably altered under the Paris Agreement (PA) of 2015, which places much greater emphasis on the role of the private sector, civil society and voluntary approaches. Although states continue to have a range of responsibilities around monitoring and reporting of their activities, these requirements fall under the commitments pledged under the framework of each state's nationally determined contributions (NDCs) (Glynn, Cadman, & Maraseni, 2017). The 'bottom-up' philosophy of the Paris Agreement is likely to have major implications for the future of both the CDM and REDD+. Although Article 5 confirms the 'existing framework' by which REDD+ functions, as well the continuation of 'results-based payments' for emissions reduction it also calls for 'alternative policy approaches' (UNFCCC, 2015, p. 22). It is possible that this may be an oblique reference to the concept 'internationally transferred mitigation outcomes' (ITMOs) of Article 6 (6.2 and 6.3 specifically). Under Article 6.4, the Agreement appears to pave the way for the trading of ITMOs by means of what is referred to as a 'mechanism to contribute to the mitigation of greenhouse gas emissions and support sustainable development' (unofficially referred to as the 'sustainable development mechanism'—SDM). But there is no specific reference to the CDM, and its fate remains unclear (UNFCCC, 2015, p. 23).

Decoupling the SDM and REDD+ via two discrete Articles under the Agreement reflects the conflict amongst Parties over whether there should continue to be a role for voluntary market-based approaches to reducing emissions, or whether there should be greater government regulation and by means of non-market measures. Although the viability of the international carbon market continues to be elusive, there continue to be Parties who are pushing for one, despite the ongoing lack of certainty that the Paris Agreement will provide for one. There is obviously a market linkage underpinning the push for the SDM. However, there is still no agreement as to whether it will complement, or completely replace, the CDM (CIFOR, 2017). What is clear is that all Parties continue to support REDD+. The division between Articles 5 and 6 allows the possibility for an international mechanism for forest-based carbon offsets, while not specifically alluding to them. It is likely that future negotiations will focus on whether whatever mechanism emerges will simply focus on REDD+ projects and payments on the basis of their results, or whether there will be a space for other initiatives that will also lead to some form of offsets (GRET, 2016).

While REDD+ has its detractors, particularly amongst environmental NGOs, it seems to have avoided some of the hostility directed to the CDM historically. The CDM had a number of major design faults in its early days, notably the exploitation of the 'potency' of certain gasses by refrigerator factories in China, who switched technologies and were able to sell their resulting hydrofluorocarbon savings as offsets for extremely high prices. Of equal concern was the accreditation of a large number of dam construction projects for hydroelectric power, which proceeded largely without the full, prior and informed consent (FPIC) of local and indigenous communities and neglected to put in place effective benefit-sharing arrangements with affected Parties (Cadman et al., 2015, pp. 29–30). Civil society organisations strenuously opposed many of these activities, but it took the CDM until 2014, two years after the official end of the first Kyoto Protocol commitment period, to provide guidance on stakeholder consultation under the mechanism (Cadman et al., 2015, p. 50). It might be possible to conclude that the focus on safeguards in REDD+ reflects a degree of policy learning from the early days of the CDM and a desire to ensure better stakeholder participation and coordination (Fujisaki, Hyakumura, Scheyvens, & Cadman, 2016). REDD+ has nevertheless been unable to avoid criticism entirely, as the state sovereignty provisions of its various agreements continue to create tensions with non-state actors concerning unresolved issues around land tenure and indigenous rights. In addition, there is the even greater uncertainty as to whether the programme is even capable of contributing meaningfully to keeping global temperatures within the 1.5–2 °C target specified under Paris (Chokkalingam & Vanniarachchy, 2011; McAfee, 2016).

4   Key Approaches and Method

While 'governance' itself may be seen as a neutral term concerning the structures and processes used for steering and coordinating stakeholder interactions within institutions, 'good' governance has been linked to a wide range of moral, ethical and normative values. The 'thicker', or more comprehensive these values, it is argued, the greater the effectiveness of the climate regime as an integrity system, and the more substantial its institutional integrity and overall legitimacy (Breakey, Cadman, & Sampford, 2017). 'Institutional integrity' in this context refers to the ongoing capability of the institution to reliably live up to its publicly stated ideals and goals—its 'Public Institutional Justification' (Breakey, Cadman, & Sampford, 2015). By improving the depth and quality of the interaction between institutional actors and stakeholders in the wider environment, the achievement of thick governance values attends to a lynchpin part (the 'context integrity') of an institution's pursuit of integrity (Breakey et al., 2015, 2017). In developing a comprehensive set of governance values, and understanding their relationship one to another for evaluation purposes, the researchers have synthesised the work of scholars from the fields of international relations, public administration, comparative politics and environmental policy (Cadman,

2011, pp. 5–18). The researchers particularly acknowledge the pioneering work of Lammerts van Bueren and Blom (1997).

In order to understand how the various institutional arrangements for good governance relate one to another, a hierarchically consistent framework of principles, criteria and indicators (PC&I), drawn from the literature, was used. Consistency allows for the appropriate location of elements within the framework, avoiding overlap or duplication at another level and enabling a 'top-down' analysis from principles to criteria and thence to indicators. A <u>principle</u> expresses a specific value. *Criteria* function at the next level down and categorise aspects of a principle. Like principles, criteria are not measured directly but are used to determine the degree of compliance with the principle. They are connected to **indicators** that are hierarchically lower and denote quantitative or qualitative parameters. Indicators (as they relate to the relevant criterion and larger principle) can then be used for measurement—in this case measures of the quality of governance. Standards are a set of PC&I that serve as a basis for monitoring and reporting, or as a reference for assessment of conditions 'on the ground'—in this case, how governance is expressed at any institutional level (Lammerts van Bueren & Blom, 1997, pp. 5–35). Table 1 sets out the hierarchical relationship between these PC&I.

This framework was used as the basis for a series of Internet surveys, conducted between 2010 and 2015, of stakeholders participating in REDD+ and CDM. See Table 2.

Survey participants were recruited through Internet-searches of publicly available participants' lists relevant to UNFCCC, REDD+ and CDM. The ensuing email addresses were entered into a database, and the online survey tool SurveyMonkey (http://www.surveymonkey.net) was used to manage the survey. Survey participants were asked to rate the governance quality of REDD+ and the CDM. Survey respondents rated their perceptions of the governance quality by means of a five-point Likert scale, using the terms 'very low', 'low', 'medium', 'high' and 'very high'. Participants were sent a survey and provided the option of clicking on a link, which took them to the survey, or they could select an option to remove themselves from the list. In addition to the Likert-scale, respondents were invited to make substantive comments relevant

Table 1 Hierarchical framework for the assessment of governance quality (Cadman, 2011: 17)

Principle (level 1)	Criterion (level 2)	Indicator (level 3)
'Meaningful participation'	Interest representation	Inclusiveness
		Equality
		Resources
	Organisational responsibility	Accountability
		Transparency
'Productive deliberation'	Decision making	Democracy
		Agreement
		Dispute settlement
	Implementation	Behavioural change
		Problem solving
		Durability

Source Cadman (2009, p. 104). *Note* Text format denotes hierarchical level (Principle, *Criterion*, **Indicator**)

Table 2 Summary of survey questions

Question	Relevant indicator
Do you think [element] is inclusive of your interests?	Inclusiveness
Do you think [element] treats all interests equally?	Equality
What level of resources does [element] provide for you to participate?	Resources
Do you think the various institutional elements in which you participate are accountable in their dealings with you regarding the [element] process?	Accountability
Do you think the various institutional elements in which you participate are transparent in their dealings with you regarding the [element] process?	Transparency
Do you consider the [element] processes in which you participate to act in a democratic manner?	Democracy
Do you consider the making of agreements in [element] to be effective?	Agreement
Do you consider the settling of disputes in [element] to be effective?	Dispute settlement
Do you think [element] will contribute to changing the behaviour that leads to deforestation and forest degradation in developing countries?	Behavioural change
Do you think [element] will help solve the problem of deforestation and forest degradation in developing countries?	Problem solving
Do you consider [element] will be durable?	Durability

Source Cadman and Maraseni (2013, p. 218) and Cadman, Maraseni, Ma, and Lopez-Casero (2017, p. 11). *Note* Introductory materials and explanatory text not included to each indicator and asked if they wished to be interviewed. The surveys were deployed for one month, with three (weekly) reminder emails, and a final forty-eight-hour closure notice. The surveys were anonymous, with confidentiality assured, but with the option for the lead researcher to identify individual respondents.

The survey cohorts (and respondents) were comprised largely of members of the respective policy communities. Respondents were asked to identify as 'Environmental', 'Social', 'Economic', 'Government', 'Secretariat or other institutional component' and 'Other' (who were then asked to specify further). Typically, these were government officials, representatives of intergovernmental and non-governmental organisations (environmental, social and economic), members of the elements under investigation (secretariats or other institutional components) and a few 'others', notably academics and researchers, and individuals who chose to represent themselves more specifically as 'other' (e.g. 'private sector' or 'indigenous peoples' organisation'). Respondents were also invited to identify whether they came from the 'Global North' or 'Developed country', and 'Global South' or 'Developing Country'. 'Environment' and 'government' were generally the highest participating groups, followed by 'other', and with a smattering of 'secretariat', followed by 'social' and 'economic'. Generally speaking, North and South were relatively equally represented, with small numbers fluctuating either way in each of the surveys.

Two initial surveys were deployed in 2010 for REDD+ (March) and the CDM (November), which were sent to approximately 800 email addresses for each mechanism. These links proved difficult for tracking respondents (especially if individuals forwarded the link). Subsequently, a unique link for each individual respondent was used, generated by the online survey tool and sent via email. The response rates from these surveys were very low, between 3 and 4%, although it should be noted that online surveys have generally lower participation rates than other forms of survey technique (Van Selm & Jankowski, 2006). Further addresses were added to the original survey cohorts to increase the number of responses, approximately 1000 REDD+ and 1300 for CDM (accounting for defunct emails and opting out). In the 2010 surveys, completion rates were lower than in 2012 (45 out of 72 for CDM or 63%; and 44 out of 164 for REDD+ or 27%). Subsequent to implementation of individual emails in 2012, completion rates were much higher at around 95%, although the overall number of respondents was lower (32 for CDM and 36 for REDD+). These numbers were not high enough to generate anything other than largely anecdotal results, particularly as the respondents across the two mechanisms were not homogenous, which the authors acknowledge, but the results, discussed below, are nevertheless interesting.

In 2014, recognising the need to refresh the existing databases of respondents, given the inevitable changes in the policy community over time, and to increase response rates, the researchers collected further emails of participants active in a range of climate and sustainability policy mechanisms, including CDM and REDD+, and combined them into a single cohort of approximately 5300 potential respondents (accounting for defunct emails and opting out). A further category 'academic' was added to this survey on account of the number of 'other' respondents, who specified this sector from the 2010 to 2012 surveys when asked to clarify. This survey was deployed in February 2015, generating 107 completed responses out of 108 attempts: 90 of these respondents opted to evaluate CDM and REDD+. In the case of REDD+, 41 identified as 'Global North' (developed countries), 49 'Global South' (developing country); for CDM, 38 were from the North, 52 from the South. Again, it should be stressed that the respondents were not identical across mechanisms in the 2015 survey, but they were from the same initial pool of 107 respondents compared to the 2010/2010 REDD+ and CDM respondents, which were completely different from each other.

The combined results of these surveys are included in Table 3. The results from the 2015 survey are further presented by region (North/South) in Table 4. For assessment purposes, each responding sector was equally weighted to avoid numerical bias; this same approach was adopted for evaluation by region.

5   Key Results

See Tables 3 and 4.

Table 3 REDD+ and CDM quality of governance (2010–2015), all sectors combined (weighted average)

Element. Date. Number	Principle: Meaningful participation (Maximum score: 25; minimum score 5)								Principle: Productive deliberation (Maximum score: 30; minimum score: 6)									Total (out of 55)
	Criterion: Interest representation (Maximum score: 15; minimum score: 3)				Criterion: Organisational responsibility (Max: 10; min: 2)			Principle Score	Criterion: Decision-making (Maximum score: 15; minimum score: 5)				Criterion: Implementation (Maximum score: 15; minimum score: 3)				Principle Score	
	Indicators				Indicators				Indicators				Indicators					
	Inclusive.	Equality	Resources	Score	Account.	Trans.	Score		Demo.	Agree.	Dispute	Score	Behaviour	Problem	Durability	Score		
REDD+ Feb. 15 (90)	3.3	3.0	2.2	8.5	3.0	3.0	6	14.5	2.8	3.0	2.8	8.6	3.1	3.1	3.2	9.4	18	32.5
REDD+ Oct. 2012 (36)	3.3	3.0	2.2	8.5	3.0	2.9	5.9	14.4	3.0	2.9	2.8	8.7	3.1	2.9	3.0	9	17.7	32.1
REDD+ Mar. 2010 (38)	3.8	3.4	2.0	9.2	3.1	3.4	6.5	15.7	3.4	2.8	2.8	9	3.7	3.5	3.8	11	20	35.7
CDM Feb. 2015 (90)	2.8	2.7	2.0	7.5	3.0	2.9	5.9	13.4	2.7	2.9	2.7	8.3	2.8	2.7	2.8	8.3	16.6	30
CDM Nov. 2012 (32)	3.6	3.0	2.4	9	3.5	3.3	6.8	15.8	3.2	3.3	3.2	9.7	3.5	3.1	2.8	9.4	19.1	34.9
CDM Nov. 2010 (45)	3.0	2.3	2.4	7.7	2.7	2.7	5.4	13.1	2.7	2.9	2.6	8.2	2.8	2.6	2.8	8.2	16.4	29.5
REDD Overall Average	3.5	3.1	2.1	8.7	3.1	3.1	6.2	14.9	3.1	2.9	2.8	8.8	3.3	3.2	3.3	9.8	18.6	33.5
CDM Overall Average	3.1	2.7	2.3	8.1	3.1	3.0	6.1	14.2	2.9	3.0	2.8	8.7	3.0	2.8	2.8	8.6	17.3	31.5

Table 4 REDD+ and CDM quality of governance (2015), North and South (weighted average)

Element, Date, Number	Indicators							Principle Score	Indicators				Indicators				Principle Score	Total (out of 55)
	Inclusive.	Equality	Resources	Criterion score	Account.	Trans.	Criterion score		Demo.	Agree.	Dispute	Criterion score	Behaviour	Problem	Durability	Criterion score		
REDD+ Feb. 2015 North (41)	3.7	2.9	**1.6**	8.1	2.8	2.7	5.6	13.7	2.5	2.8	2.5	7.7	3.1	2.9	3.1	9.1	16.9	30.5
REDD+ Feb. 2015 South (49)	3.1	3.1	2.8	9.0	3.4	3.2	6.6	15.6	3.1	3.2	3.0	9.3	3.1	3.2	3.2	9.5	18.8	34.4
CDM Feb. 2015 North (38)	3.0	2.7	**1.6**	7.3	2.7	2.7	5.3	12.6	**2.4**	2.8	**2.3**	7.4	2.7	**2.4**	2.6	7.8	15.2	27.8
CDM Feb. 2015 South (52)	2.7	2.7	**2.4**	7.8	3.2	3.1	6.3	14.1	2.9	3.0	3.0	8.9	2.8	2.9	2.9	8.6	17.4	31.5

Source Cadman et al. (2015) and Maraseni and Cadman (2015)

Notes (1) Light-grey represents the highest scoring indicators; (2) Medium-grey the lowest scoring indicators; and (3) Numbers in **bold** are a 'fail' at the indicator level

5.1   Commentary—2015, 2012 and 2010 Surveys of REDD+ and CDM—Combined Results

In the 2015 survey, REDD+ received a high 'pass'. **Inclusiveness** was the highest rated indicator, and **resources** the lowest (a 'fail'). The results from the 2012 survey—despite being from a largely different cohort of respondents—are remarkably similar to 2015: a slightly weaker 'pass', again with **inclusiveness** the highest rated indicator and **resources** the lowest (both exactly the same scores as the 2015 survey). The 2010 survey generated the most positive results: a low 'credit' (64%); the highest rating indicator was again **inclusiveness** but this time, shared with **durability** (3.8—the highest two ratings across all surveys). The lowest indicator was once again **resources** (another 'fail'). In the 2015 survey, CDM received a low 'pass'. This time, **accountability** was the highest rated indicator, with **resources** again the lowest (and a 'fail'). The 2012 survey generated the most positive results for the mechanism, with a high 'pass' (64%). **Inclusiveness** was the highest rated indicator, **resources** the lowest (another 'fail'). The 2010 survey was similar to that of 2015: a low 'pass', with **inclusiveness** the highest rating indicator, but **resources** being outranked as the worst indicator by **equality** (2.3 cf. 2.4—both 'fails'). Overall, REDD+ clearly performed better than CDM, although interestingly, while REDD+ received the higher rating for **inclusiveness** (3.5 cf. 3.1), CDM had a slightly better rating for **resources** (2.1 cf. 2.3—both 'fails' nevertheless).

5.2   Commentary—2015 Surveys of REDD+ and CDM—North and South

REDD+ received a relatively high 'pass', but not a 'credit'. The mechanism did pass all principles and criteria across Northern and Southern respondents, however, with Southern respondents providing consistently higher ratings. **Resources** was again the weakest indicator, with Northern respondents providing a much lower rating than their Southern counterparts (1.6, a 'fail' cf. 2.8). Interestingly, the indicator ratings provided by Northern respondents were more dynamic than those from the South and included the highest rated indicator (**inclusiveness**). With the exception of **resources**, Southern participants rated REDD+ governance indicators within the 'high' band, with little differentiation. For the South, **accountability** was the highest performing indicator. Respondents did not appear to be especially impressed with the CDM, however, which received a low 'pass' only. Southern respondents rated the mechanism more favourably—a result that was repeated across both principles. Northern respondents identified **interest representation** and **decision-making** as the weakest criteria. At the indicator level, both Southern and Northern respondents gave **resources** a low—'fail'— score, and for Northern respondents, this was the lowest scoring indicator. **Democracy** and **dispute settlement** also received a low rating from the North (both 'fails'). The highest performing indicators were **inclusiveness** (North) and **agreement** and **dispute settlement** (South). These higher ratings for **agreement** and **dispute settlement** are offset by a 'fail' for **dispute settlement** amongst Northern respondents (a 'fail').

6   Conclusions

The relative consistency of results for REDD+ and CDM across years and indicators is remarkable, particularly given the different sets of respondents and response rates. With all scores exceeding thirty, and some exceeding thirty-five, it is a fair observation to say that overall, the perceptions of governance quality of REDD+ amongst respondents were positive. However, there were some inconsistencies in perceptions of governance quality between Northern and Southern respondents in the 2015 survey, with Northern respondents giving a considerably lower score than their Southern counterparts. CDM performed consistently more weakly than REDD+. The highest-scoring indicator was generally 'inclusiveness', a positive sign, given the multi-stakeholder nature of climate governance (even if the state remains dominant). Interestingly, there was some recognition given to CDM for its accountability (perhaps reflecting a degree of confidence in its robust accounting methods). However, it is alarming to note that, across years and elements, the lowest scoring indicator was 'resources'—the only exception being 'equality' for the 2010 CDM survey (which may reflect the generally poorer perception of CDM amongst respondents in comparison with REDD+).

Given that resources, or capacity, are essential for interest representation, and meaningful participation, attention must be given to addressing this critical area of good governance, as inclusiveness is not in itself sufficient for ensuring good interest representation. Stakeholders require the capacity (financial, educational, technical, institutional and so on) to ensure that their participation is meaningful, rather than tokenistic. In the planning, implementation and evaluation of mechanisms for combatting climate change, equitable allocation of resources for participation—both in policy development and in implementation—will be one of the principal challenges. This conclusion also informs the recent 'procedural turn' in the literature on climate justice, showing how vital resource allocations are to secure the ideals laid down by deliberative justice. Participants need not only formal inclusion in deliberations about norms and policies, but the material and technical capabilities to make a genuine contribution. Resources are also key in terms of implementation. Without improving livelihoods in a sustainable way, indigenous and local communities may prefer to convert forests to other land uses, and developing states may then struggle to meet their NDC targets. However, if the indications of developing state respondents in the 2015 survey are anything to go by, there may be a greater level of support for, and confidence in, the merits of such mechanisms in the Global South, than in the Global North.

It should be further noted that the results were not entirely negative for CDM. Respondents felt that it had relatively good accountability mechanisms in place, although not markedly better than REDD+ (perhaps the latter has learned from CDM with its carbon accounting measures, although this remains to be seen). It is also worth noting that Northern respondents generally provided lower ratings for everything than Southern respondents, reflected in both CDM and REDD+. The consistently high ratings for inclusiveness compared to the low ratings for resources would appear to suggest that participants in both REDD+ and CDM felt that they had relatively good access to these mechanisms, but that the allocation of resources to facilitate participation is inadequate. The caveat should be added here that the surveys are capturing people on these forums' participants' lists, so they are by definition included at least to that extent. The value of measuring their perceptions nevertheless is that it allows a comparison between the two. The conclusion suggested is that REDD+ is more inclusive than CDM.

If the SDM is to be successful, it needs to learn from the lessons of both REDD+ and the CDM and pay greater attention to capacity building amongst under-resourced stakeholders. As these surveys have shown, the application of standards covering a broad suite of governance values, and not just focusing narrowly on accountability and transparency, should be seriously considered by donor states and private sector investors, if they wish to ensure the integrity of climate governance and its proper financing. It should give further attention to allocating resources to increase participation and in sharing future benefits from reducing emissions in the effort to prevent dangerous climate change. Finally, the SDM should give thought to the underlying tensions between developed and developing state participants in climate governance. These could prove damaging in future if they are not addressed immediately.

References

Abreu Mejía, D. (2010). The evolution of the climate change regime: Beyond a north-south divide?

Allan, J. I., & Dauvergne, P. (2013). The global south in environmental negotiations: The politics of coalitions in REDD+. *Third World Quarterly, 34*(8), 1307–1322.

Bäckstrand, K., & Lövbrand, E. (2007). Climate governance beyond 2012: Competing discourses of green governmentality, ecological modernization and civic environmentalism. In *The social construction of climate change: Power, knowledge, norms, discourses* (pp. 123–147). Hampshire: Ashgate.

Boran, I. (2016). Principles of public reason in the UNFCCC: Rethinking the Equity Framework. *Science and Engineering Ethics, 23*(5), 1253–1271. https://doi.org/10.1007/s11948-016-9779-9.

Boran, I., & Katz, C. (2017). Climate change justice. In *Routledge encyclopedia of philosophy: Taylor and Francis.* Retrieved from https://www-rep-routledge-com.libraryproxy.griffith.edu.au/articles/thematic/climate-change-justice/v-1.

Breakey, H. (2015). COP20's ethical fallout: The perils of principles without dialogue. *Ethics, Policy & Environment, 18*(2), 156–169.

Breakey, H., Cadman, T., & Sampford, C. (2015). Conceptualizing personal and institutional integrity: The comprehensive integrity framework. *Research in Ethical Issues in Organizations, 14,* 1–40. https://doi.org/10.1108/S1529-209620150000014001.

Breakey, H., Cadman, T., & Sampford, C. (2017). Governance values and institutional integrity. In T. Cadman, R. Maguire, & C. Sampford (Eds.), *Governing the climate change regime: Institutional integrity and integrity systems* (pp. 16–44). Abingdon: Routledge.

Cadman, T. (2009). *Quality, legitimacy and global governance: A comparative analysis of four forest institutions.* Doctoral thesis, University of Tasmania, Launceston.

Cadman, T. (2011). *Quality and legitimacy of global governance: Case lessons from forestry.* New York and Basingstoke: Palgrave Macmillan.

Cadman, T. (2013). Introduction: Global governance and climate change. In T. Cadman (Ed.), *Climate change and global policy regimes* (pp. 1–16). London: Palgrave Macmillan.

Cadman, T., & Maraseni, T. (2013). More equal than others? A comparative analysis of state and non-state perceptions of interest representation and decision-making in REDD+ negotiations. *Innovation.* https://doi.org/10.1080/13511610.2013.771880.

Cadman, T., Eastwood, L., Michaelis, F. L.-C., Maraseni, T. N., Pittock, J., & Sarker, T. (2015). *The political economy of sustainable development: Policy instruments and market mechanisms.* Cheltenham: Edward Elgar.

Cadman, T., Maraseni, T., Breakey, H., López-Casero, F., & Ma, H. O. (2016). Governance values in the climate change regime: Stakeholder perceptions of redd+ legitimacy at the national level. *Forests, 7*(10), 212.

Cadman, T., Maraseni, T., Ma, H. O., & Lopez-Casero, F. (2017). Five years of REDD+ governance: The use of market mechanisms as a response to anthropogenic climate change. *Forest Policy and Economics, 79,* 8–16. https://doi.org/10.1016/j.forpol.2016.03.008.

CDM-Policy-Dialogue. (2012a). *Climate change, carbon markets and the CDM: A call to action.* Retrieved from Luxembourg http://www.cdmpolicydialogue.org/report/rpt110912.pdf.

CDM-Policy-Dialogue. (2012b). *High-level panel on the CDM Policy Dialogue.* Retrieved from http://cdmpolicydialogue.org.

Chapman, S., Maguire, R., Doshi, M., Kago, C. W., Aquino, N. K., Kiguatha, L., ..., Engbring, G. (2015). The elements of benefit-sharing for REDD+ in Kenya: A legal perspective. *Carbon & Climate Law Review: CCLR, 9*(4), 283.

Chapman, S., Wilder, M., & Millar, I. (2014). Defining the legal elements of benefit sharing in the context of REDD. *Carbon & Climate Law Review: CCLR, 8*(4), 270–281.

Chapman, S., Wilder, M., Millar, I., Dibley, A., Yeang, D., Heffernan, J., ..., Dooley, E. (2015). A legal perspective of carbon rights and benefit sharing under REDD+: A conceptual framework and examples from Cambodia and Kenya. *Carbon & Climate Law Review : CCLR, 9*(2), 143.

Chokkalingam, U., & Vanniarachchy, A. (2011). *Sri Lanka's REDD+ potential, myth or reality?* Forest Carbon Asia Colombo.

CIFOR. (2017). *Analysis: Getting down to business in Bonn.* Retrieved from https://forestsnews.cifor.org/49709/getting-down-to-business-in-bonn?fnl=en.

de Oliveira, J. P., Cadman, T., Ma, H. O., Maraseni, T., Koli, A., Jadhav, Y. D., et al. (2013). *Governing the forests: An institutional analysis of REDD+ and community forest management in Asia.* Yokohama: International Tropical Timber Organization (ITTO)/United Nations University Institute of Advanced Studies (UNU-IAS).

Dooley, K., & Gupta, A. (2017). Governing by expertise: The contested politics of (accounting for) land-based mitigation in a new climate agreement. *International Environmental Agreements: Politics, Law and Economics, 17*(4), 483–500.

FCPF. (2011). Forest carbon partnership facility common approach to environmental and social safeguards for multiple delivery partners.

FCPF. (2017). *About FCPF.* Retrieved from https://www.forestcarbonpartnership.org/about-fcpf-0.

Forsyth, T. (2009). Multilevel, multiactor governance in REDD+: Participation, integration and coordination. In A. Angelsen (Ed.), *Realising REDD+: National strategy and policy options* (pp. 113–122). Bogor: Center for International Forestry Research (CIFOR).

Fujisaki, T., Hyakumura, K., Scheyvens, H., & Cadman, T. (2016). Does REDD+ ensure sectoral coordination and stakeholder participation? A comparative analysis of REDD+ national governance structures in countries of Asia-Pacific region. *Forests, 7*(9), 195.

Glynn, P. J., Cadman, T., & Maraseni, T. N. (2017). *Business, organized labour and climate policy: Forging a role at the negotiating table.* Cheltenham: Edward Elgar.

GRET. (2016). *What role do carbon markets play in the Paris Agreement?* Retrieved from http://www.gret.org/discover-gret/about-us/?lang=en.

Harada, K., Prabowo, D., Aliadi, A., Ichihara, J., & Ma, H.-O. (2015). How can social safeguards of REDD+ function effectively conserve forests and improve local livelihoods? A case from Meru Betiri National Park, East Java. *Indonesia. Land, 4*(1), 119–139. https://doi.org/10.3390/land4010119.

Hoffmann, M. J. (2011). *Climate governance at the crossroads: Experimenting with a global response after Kyoto.* Oxford: Oxford University Press.

Jänicke, M. (1992). Conditions for environmental policy success: An international comparison. *The Environmentalist, 12,* 47–58. https://doi.org/10.1007/BF01267594.

Jänicke, M. (Ed.). (1996). *Democracy as a condition for environmental policy success: The importance of non-institutional factors.* Cheltenham and Lyme: Edward Elgar.

Knieling, J., & Leal Filho, W. (2012). *Climate change governance.* Heidelberg: Springer Science & Business Media.

Koenig-Archibugi, M. (2006). Introduction: Institutional diversity in global governance. In M. Koenig-Archibugi & M. Zurn (Eds.), *New modes of governance in the global system: Exploring publicness, delegation and inclusiveness* (pp. 1–30). Basingstoke: Palgrave Macmillan.

Lammerts van Beuren, E. M., & Blom, E. M. (1997). *Hierarchical framework for the formulation of sustainable forest management standards.* Leiden: The Tropenbos Foundation.

Lederer, M. (2011). From CDM to REDD+—What do we know for setting up effective and legitimate carbon governance? *Ecological Economics, 70*(11), 1900–1907. https://doi.org/10.1016/j.ecolecon.2011.02.003.

Lyster, R. (2011). REDD+, transparency, participation and resource rights: The role of law. *Environmental Science & Policy, 14*(2), 118–126.

Maguire, R. (2015). Mapping the integrity of differential obligations within the United Nations framework convention on climate change. In V. Popovski, R. Maguire, & H. Breakey (Eds.), *Ethical values and the integrity of the climate change regime* (pp. 31–42). Farnham: Ashgate.

Maraseni, T. N., & Cadman, T. (2015). A comparative analysis of global stakeholders' perceptions of the governance quality of the clean development mechanism (CDM) and reducing emissions from deforestation and

forest degradation (REDD+). *International Journal of Environmental Studies,*
72(2), 288–304. https://doi.org/10.1080/00207233.2014.993569.

Mason, M. (1999). *Environmental democracy.* New York: St. Martin's Press.

McAfee, K. (2016). Green economy and carbon markets for conservation
and development: A critical view. *International Environmental Agreements:*
Politics, Law and Economics, 16(3), 333–353.

McGregor, A., Weaver, S., Challies, E., Howson, P., Astuti, R., & Haalboom,
B. (2014). Practical critique: Bridging the gap between critical and prac-
tice-oriented REDD+ research communities. *Asia Pacific Viewpoint, 55*(3),
277–291. https://doi.org/10.1111/apv.12064.

Newell, P. (2009). Varieties of CDM governance: Some reflections. *The Journal*
of Environment & Development, 18(4), 425–435.

Radunsky, K., Cadman, T. (2017). Afterword: The long road to Paris: Insider
and outsider perspectives. In R. Maguire, C. Sampford, T. Cadman (Eds.),
Governing the climate change regime: Instituional integrity and integrity systems
(pp. 250–265). Abingdon: Routledge.

Rhodes, R. A. W. (1997). *Understanding governance: Policy networks, govern-*
ance, reflexivity and accountability. Buckingham: Open University Press.

Roberts, J. T., & Parks, B. (2006). *A climate of injustice: Global inequality,*
north-south politics, and climate policy. Cambridge, MA: MIT Press.

Schmidt, V. A. (2013). Democracy and legitimacy in the European Union
revisited: Input, output and 'throughput'. *Political Studies, 61,* 2–22.

Scholte, J. A. (2004). Civil society and democratically accountable global
governance. *Government and Opposition, 39*(2), 211–233. https://doi.
org/10.1111/j.1477-7053.2004.00121.x.

Smismans, S. (2004). *Law, legitimacy, and European governance.* Oxford:
Oxford University Press.

Stiglitz, J. E. (2003). Globalization and development. In D. Held &
M. Koenig-Archibugi (Eds.), *Taming globalisation: Frontiers of governance*
(pp. 47–67). Cambridge: Polity Press.

Thompson, M. C., Baruah, M., & Carr, E. R. (2011). Seeing REDD+ as a
project of environmental governance. *Environmental Science & Policy, 14*(2),
100–110.

UNFCCC. (2011). *Outcome of the work of the ad hoc working group on long-*
term cooperative action under the convention (Cancun Agreements). Retrieved
from https://unfccc.int/resource/docs/2010/cop16/eng/07a01.pdf.

UNFCCC. (2014). *Warsaw framework for REDD-plus.* Retrieved from http://
unfccc.int/land_use_and_climate_change/redd/items/8180.php.

UNFCCC. (2015). *Paris Agreement as contained in the report of the conference of the parties on its twenty-first session.* Retrieved from http://unfccc.int/files/home/application/pdf/paris_agreement.pdf.

UN-REDD. (2009, October 29–30). *Report of the third policy board meeting Washington, DC.* Retrieved from http://www.unredd.net/index.php?option=com_docman&Itemid=134&view=document&alias=1234-final-report-of-the-3rd-policy-board-english-1234&category_slug=report-3rd-policy-board-meeting-eng-fr-sp-433.

UN-REDD. (2012). *UN-REDD programme social and environmental principles and criteria.*

Van Selm, M., & Jankowski, N. W. (2006). Conducting online surveys. *Quality & Quantity, 40*(3), 435–456.

Young, I. M. (2000). *Inclusion and democracy.* Oxford: Oxford University Press.

11

Constraining Supply: The Moral Case for Limiting Fossil Fuel Exports

Jeremy Moss

1   Introduction

Since the first Intergovernmental Panel on Climate Change report in 1990, most of the world's states have committed to reduce their greenhouse gas emissions to avert the threat of dangerous climate change. Yet their promised commitments and their actual reductions have not been a success. The measures that they have chosen have failed to produce the necessary rate of emission reductions (Victor et al., 2017). With each year that passes the problem becomes more difficult and costly to solve (Stern, 2007). This situation raises the question of what policies states ought now to adopt to remedy this failure. One important dimension of this debate is whether states should adopt policies that reduce the demand for fossil fuels—through carbon consumption taxes for example—or through constraining the supply of fossil fuels. How

J. Moss (✉)
University of New South Wales, Sydney, NSW, Australia
e-mail: j.moss@unsw.edu.au

© The Author(s) 2019

B. Edmondson and S. Levy (eds.), *Transformative Climates and Accountable Governance*, Palgrave Studies in Environmental Transformation, Transition and Accountability, https://doi.org/10.1007/978-3-319-97400-2_11

we evaluate these different measures will depend on a range of factors such as their efficiency, fairness and feasibility. I want to focus particularly on the moral dimension of this evaluation and argue that there is a strong prima facie case for restricting the supply of fossil fuels through restricting exports ('supply side' constraints) as it best accounts for the moral injunction not to contribute to harm through increasing greenhouse gas emissions.

Utilising some greenhouse gas emissions figures from Australia as illustrative examples, this chapter argues that many fossil fuel-exporting states play major and morally culpable roles in contributing to global emissions. Because of the contribution to the risk of harm that exporting states make, we need to be clear about what moral responsibility they have for their contribution and what actions they ought to take in response. The question that this chapter poses is: If there is a moral case to constrain exports, what should the consequences be for those who continue to export? This question has added significance given there is currently little or no global governance of exports and it would be desirable to create a framework to govern the export of fossil fuels and the consequences that should follow.

2   The Export Context

Many states are significant exporters of fossil fuels. For some states, such as Saudi Arabia or Norway, it is or has been a major source of their wealth. In 2015/2016, Australia exported 388 megatonnes of coal, more than any other state and representing 30% of world trade in coal (Australian Government, 2016). The value of these exports was A$34 billion and directly employed 44,000 people within Australia (Australian Government, 2016). The picture is similar for Australia's other large fossil fuel export—gas. Australia is currently the second-largest exporter of gas (Cassidy & Kosev, 2015). By 2021, it will be the largest exporter of gas in the world, overtaking Qatar (Cassidy & Kosev, 2015). This development follows investment of approximately A$200 billion into the industry over the last decade (Dediu et al., 2016). In dollar terms, gas is expected to become Australia's second-largest commodity export (after iron ore) by 2018 (Cassidy & Kosev, 2015).

The volume of coal and gas exported from Australia and the greenhouse gas emissions they produce are significant relative to Australia's domestic production of emissions. Table 1 summarises past, present and projected figures for total Australian CO_2-e emissions (exports excluded), CO_2-e emissions from exported coal, CO_2-e emissions from exported gas and also details Australia's total emissions target for 2020.[1] As shown in Table 1, Australia's exported coal and gas have contributed approximately twice the volume of emissions, relative to Australian-based emissions in recent years. Projections for 2018–2021 suggest that emissions from exported coal and gas will continue to increase by modest margins while anticipated domestic emissions are expected to stabilise over this timeframe. In spite of a 2020 Australian emissions target of 532 megatonnes, current projections of total emissions suggest a likely output of closer to 2000 megatonnes (with approximately half of this deriving from exported coal and gas).

Australia also has huge remaining coal and gas reserves, with a conservative estimate of carbon dioxide CO_2-e emissions associated with such reserves sitting at 14,419.07 megatonnes.[2] The scale of these reserves and their actual and projected emissions raise important moral questions concerning whether and to what extent exporting countries ought to share liability for the increased risk of dangerous climate change.

3   Exporting Harm

With the exception of air and sea travel, most of the greenhouse gases are produced by people within the territory of states, so when it comes to assigning responsibility for emissions and the harm to

---

[1] All figures are in Mt CO_2-equivalent (Mt CO_2-e). Australian domestic emissions figures are taken from the Department of the Environment's Australia's Emission Projections 2029–2030 (Department of the Environment, 2015, p. 32). Export figures are taken from the Office of the Chief Economist, Resources and Energy Quarterly (Office of the Chief Economist, 2016, p. 79). Conversion factors used are given in the Australian Government's Greenhouse Gas Accounts Factors (Australian Government, 2016, p. 12).

[2] The gas reserves figure is based on GeoScience Australia's estimates of total identified gas resources in Australia (2016). Conversion factors used are given in the Australian Government's Greenhouse Gas Accounts Factors (August 2016, p. 12).

Table 1 Australian actual and projected emissions including fossil fuel exports

	Unit	2014–2015	2015–2016	2016–2017	2017–2018	2018–2019	2019–2020	2020–2021
Total Australian emissions of Co_2-e with exports excluded	Mt CO2-e	565.6	594.4	617.3	638.2	652.7	655.6	659.8
Total Australian emissions from coal exports	Mt CO_2-e	1016.36	1011.92	1012.52	1032.68	1048.6	1063.69	1074.44
Total Australian emissions from natural gas exports	Mt CO_2-e	67	95.95	135.88	182.51	198.32	199.13	201.54
Total actual emissions	Mt CO_2-e	1648.96	1702.27	1765.7	1853.39	1899.62	1918.42	1935.78
Total Australian emissions 2020 target								532

which they might have contributed, the task ought to be fairly simple: states ought to take responsibility for those emissions and any associated harms, to which they have contributed. This kind of approach is utilised by the United Nations Framework Convention on Climate Change (UNFCCC) when it determines national (domestic) inventories of greenhouse gas emissions as part of the global carbon budget. The global carbon budget represents the finite amount of greenhouse gases that humankind can emit, consistent with an acceptable likely rise in global temperatures. According to the UNFCCC guidelines, 'greenhouse gas emissions and removals taking place within national territory and offshore areas over which the country has jurisdiction' (Eggleston, Buendia, Miwa, Ngara, & Tanabe, 2006, 1.4) are included within that state's carbon budget.

What the UNFCC calls 'scope 1+2' emissions are those emissions that are produced within a state's borders by various types of activity, such as industrial activity, transport and so on. Scope 3 emissions are the emissions that are produced by the commodities when they are generated outside the state's territory, and these are not part of the state's emissions budget. To give an example, if a state exports coal, the emissions that are generated by extracting the coal and transporting it to a port are part of that state's emissions budget because they occur within its territorial boundary. The emissions that are produced when the coal is burnt are part of the actual budget of the state that burns them. Including emissions in a state's budget is one way of taking responsibility for emissions, but not the only way. A state could assume financial obligations to pay the costs of adaptation or mitigation or compensate others for harms done. Nonetheless, the budget model above serves to illustrate how responsibility for emissions is determined.

If the formula set down by the UNFCCC in counting emissions is applied to liability for harm from emissions, states exporting fossil fuels are not deemed responsible for the harms that may result from the emissions produced from burning them. However, there is a plausible moral case for claiming that states have responsibility for at least part of the harms caused by their scope 3 emissions as well as a duty to stop causing those harms.

To see how this might be the case, let's bracket climate issues and consider the following analogies. Suppose that a state produced and exported large amounts of tobacco to a developing state that did not have health warnings for smoking. Given what we know about the links between smoking and death and disease, the exporting state is plausibly implicated in the harm caused and morally responsible for at least some of that harm. Another example concerns hazardous waste. Where one state knowingly ships its dangerous medical or industrial waste to a state that has low or no standards for its safe disposal, we can say that the exporter bears some responsibility for harms that may result when the waste is not disposed of properly. This is likely to be the case even where there was consent from the importer to take the waste.

More obvious still is the case of uranium exports. There are good reasons why many states place restrictions on the end destinations of their product. States place restrictions on the sale of uranium because the risks of weapons proliferation, accidents at reactors, storage issues and so on are just too great for some states to countenance export programmes. States such as Australia are bound by the Nuclear Non-Proliferation Treaty to only sell uranium for peaceful uses (UNODA, 1970). Should one state knowingly export uranium to another state where safety is lax, it could rightly be accused of being irresponsible and having a share in the blame if an accident were to happen.

What the cases above have in common is that they cause harm in morally significant and blameworthy ways. To return to the fossil fuel case, coal and gas are commodities which we know are likely to cause harm when we export them for their standard uses. We know that gas or coal exported from Australia or Brazil to China will be used to generate electricity or produce steel, and that these processes will release greenhouse gases into the atmosphere. We also know that the links between increasing greenhouse gases in the atmosphere and harms caused by them are strong. The harms that this chain of events potentially causes to the significant interests of people everywhere include: increased vulnerability to disease, crop failure, water shortages and impacts caused by severe weather events (IPCC, 2014b).

The harm caused by greenhouse gas emissions is like the harms caused by tobacco in morally relevant respects: exporting states contribute to knowingly harming the significant interests of others in ways that could, in many cases, be avoided. In wealthy states such as Norway and Australia, the export of fossil fuels is not the only means for those states to maintain a high standard of living. They may be significant to the economy, but they are not the only contributions to what is already a high standard of living. If it is the case that exported fossil fuels are exported in the knowledge that they cause significant harm or risk-causing harm to significant human interests and the practice could be avoided, then resource-exporting states have a prima facie responsibility for the harms that they cause through the export of fossil fuels. These different elements of harm (knowing, contribution, alternatives and interests) are present in the exporting case.

Let us now look in more detail at the connection between exports and the risk of harm and the question of how a state that is not directly producing the emissions might be at fault. For a state to fully accrue liability from the harms to which it contributes by selling fossil fuels, it must do so knowingly, affecting significant interests, and having alternatives. The first two conditions are relatively uncontroversial for a wealthy state such as Australia, given the sheer volume of emissions from fossil fuel exports poses significant additional climate change risks. Additionally, as a wealthy state, Australia has alternatives to generating its income in this way.

In terms of whether a state, such as Australia, had knowledge of the likely harms that might result from its exports (the knowledge condition), a plausible standard is that it must know or should know that its actions are contributing or are likely to contribute to a harm. All that is plausibly required for liability is that an agent knowingly contributes to the potentially harmful action or is ignorant (but should not have been), and that they know that their action is a contribution to the harm committed by someone else. Even if the states that emit the greenhouse gases by burning the fossil fuels intend to do so and the exporting state has no such intention, this ought not to absolve the exporter of blame because the exporting state should have known what the end use of the fossil would be. Knowing or being in a position where one should have known is a crucial part of being complicit.

Turning to the second condition, how does an exporter contribute to harm? After all, states or individuals can be involved in contributing to harmful outcomes in multiple ways. In the case of climate change, the connections between agents and those who are harmed are complex. States can contribute to causing emissions indirectly, for instance, by: supplying other emitters with fossil fuels, supporting a policy that increases emissions or prevents low-emission responses, buying products from emitters, providing subsidies to emitters, influencing the political process in favour of high-emitting companies, supplying essential resources to emitters, or stalling global climate treaties and so on.

But if the main type of contributions to climate change from exporters are indirect contributions, how are we to describe and assess the moral liability of contributing to harm in this indirect way, especially in comparison with more direct contributions? A plausible framework for explaining how exporting fossil fuels is harmful is via the language of complicity. Typically, in legal and moral theory, an agent is complicit when they knowingly assist or encourage the harms of others (Lepora & Goodin, 2013; Kutz, 2007). Through involvement in the harms done directly by primary or principal agents, a secondary agent becomes an accomplice or accessory and, according to this formula, shares some of the liability for the harms done. The badness of the harm committed by the secondary agent derives from the harm of the primary agent. A secondary agent can be complicit in different ways including: actively conspiring and planning a harm with a principal agent, simply cooperating with them, encouraging, permitting, soliciting or by aiding and abetting.

It is important to note the complexity that can be involved in being complicit in increasing the risk of climate harms through exporting fossil fuels, which involves understanding the different ways in which a state can be complicit. For example, a state could be complicit through directly *supplying* fossil fuels (being an 'upstream producer' as they are often called). The second way is through the various types of *support*. Support might take the form of providing the physical infrastructure that allows the coal and other fossil fuels to be sold: the ports, roads, rail lines and so on. A second dimension of support is the financing and subsidies that supports production. Subsidies to fossil fuel companies and products can often be difficult to categorise. Subsidies are typically thought of as being of three types: consumer, production or post-tax (externalities). Consumer subsidies typically take the form of regulating fuel prices or cash handouts to different groups of consumers. Production subsidies are directed at the fossil fuel production process itself and might include tax breaks, cheap public finance, deductions for research and development or explorations investment or indemnities for environmental hazards. Some methodologies include post-tax costs (externalities), such as the cost of pollution, which dramatically increases the estimates of the extent of fossil fuel subsidies. For instance, an IMF report incorporating externalities into its methodology estimated that in 2015 alone energy subsidies globally (mostly fossil fuels) were $5.3 *trillion* dollars (Coady, Parry, Shears & Shang, 2015). Other estimates range from $160 billion to $440 billion per year (Bast, Doukas, Pickard, van der Burg, & Whitley, 2015).

Leaving aside the merits of the different ways of calculating subsidies, what is important about each of these types of subsidies is that they provide crucial *benefits* for the fossil fuel industry, which is why subsidies are such a crucial component of support. The key feature of subsidies that ought to be of moral concern is the fact that they confer a benefit on a recipient relative to other recipients, whether the latter are individuals or specific industries. Indeed, this is how the World Trade Organisation (WTO) has understood the significance of subsidies. The WTO defines a subsidy as 'any financial contribution by a government, or agent of a government, that is recipient-specific and confers a benefit on its recipients in comparison to other market recipients'.[3] There are, in fact, a huge array of measures that might count as subsidies. A recent report by the Overseas Development Institute (Bast et al., 2015) lists 30 different types. In the case of Australia, recent reports by the IMF and Oil Change estimate that Australia spends around A$5 billion pa on support of one kind or another (Coady et al., 2015).

---

[3]WTO quoted in Overseas Development Institute and Oil Change International, Empty Promises: G20 subsidies to oil, gas and coal production' (2015, p. 32)

A plausible third category of contribution to emissions could be called *influence*. Governments, powerful lobby groups or companies themselves can influence decisions about energy policy that favour fossil fuels. This can occur via trade arrangements, fossil fuel favourable domestic legislation, securing subsidies to make coal competitive, weakening global climate agreements, providing finance for coal-related infrastructure or in general via influencing governments to favour fossil fuel production or use. Fossil fuel companies themselves often engage in this kind of influence when they want a government to adopt a particular policy.

These three ways of contributing to the risk of harm—supply, support and influence—are what makes fossil fuel-exporting states complicit. Through supplying, supporting and influencing emission outcomes, Australia is complicit in risking harming others. Of all the components that go into providing fossil fuels to a buyer, the act of selling fossil fuels is the most proximate to the harm itself. Nonetheless, without the types of indirect support listed above, it is also less likely that the fossil fuel would not have been produced or sold because the various forms of support benefit the fossil fuel industry, making their products easier to produce or sell. If states are complicit in harming others in this way via their support of exports, then it provides states with a *negative duty*—a duty not to interfere or harm others—to stop their involvement in this harm in proportion to their involvement in contributing to the harm.

In contrast, duties to act to prevent dangerous climate change might also be said to stem from the positive duties for taking action to mitigate dangerous climate change. Positive duty accounts—duties to assist others—typically base duties not on whether or not we have contributed in some way to a harm, but on whether we ought to do something positive to assist others (Singer, 1972). In the case of climate change, the duty required of wealthy states might be to provide poor states with assistance with their transition to a low-carbon economy by bearing the costs of adaptation and transition because they can more easily do so.

Of course, arguments for why states ought to act could rely on both negative and positive arguments and typically this is the case. For instance, a wealthy state might have negative duties because of its contribution to harms and positive duties because it had the capacity to alleviate the harms suffered by people in other states. One might think that because institutions are causally implicated in causing harm via their emissions, they have a duty to shoulder a greater share of the burdens, *and* that they also have positive duties to act as part of their general duty to take actions to prevent climate change. Whatever the merits of the positive duties, one could still acknowledge that harming people generates the kind of duties that we have been discussing independently of these other considerations. The point is that these are very different arguments and ought to be made clear as the strength of each is likely to be different, as is the type of objection that each must face.

Negative duty arguments for why states ought to act are stronger than positive ones because of their moral stringency. The claims that arise from violating people's significant interests are a particularly stringent kind of claim because of the weight associated with this kind of harm (Feinberg, 1984). The strength of negative duty accounts is that it is not acceptable to cause unjust harm to others in pursuit of national goals. Dumping toxic waste on another state's doorstep is not justified even if it is greatly beneficial to the dumping state. Moreover, an institution or state ought to try hard to avoid these harms. Part of the reason for this is the agency exercised in bringing about a harm. If I harm someone, then I cause them to be worse off, whereas if I merely do not assist in preventing or alleviating their harm I am arguably less morally to blame (Scheffler, 1995).

Whatever the disagreements over whether or not people or states have a duty to aid, causing a harm adds strength to any duty to rectify or compensate for harms in some way. Causing harm in an unjust way should provide a powerful and important constraint on actions (Cripps 2011; Feinberg 1984). It provides us *prima facie* reasons for an agent to restrain their actions and/or to be liable for the consequences.

Negative duty accounts are also the best way to capture what is wrong with the practice of exporting fossil fuels. This is not to say that there are not, all things considered, reasons (such as having no economic choice but to export fossil fuels) why harming someone might be permissible in certain circumstances, but that these reasons will have to be argued for. A harmful act may be perpetrated in a situation where the agent simply has no choice but to harm someone. An agent may also harm someone unintentionally through negligence or by dint of unforeseen circumstances. These and other reasons may mitigate the responsibility for the harm and its consequences. In the situation where one state is causing harm to another, as is the case with fossil fuel use and production, this generates duties to rectify or stop the harm even if there are other bases of duties. Since negative duties are such a powerful and important constraint, advocates of each position can acknowledge its force even if they also think there are other bases of duties to phase out fossil fuel production. If this is correct, it offers a way that states ought to see themselves as constrained even if there are no clearly enforceable mechanisms for accepting consequences for exported emissions. As we have noted, the Paris Climate Agreement involved states pledging to reduce their emissions, but the agreement offered little in the way of sanctions for those who do not, or a way of assessing the development, production and sale of fossil fuels for export. Understanding the contribution to harms that fossil fuel exports can make via a harm principle provides clear guidance for how states ought to self-govern their production and sale of fossil fuels.

4   Policy Implications

In the case where a state is morally complicit through supporting and supplying fossil fuel exports, what ought the consequences of that complicity be? In particular, what actions ought a state take in response to its (continuing) complicity? This is an important but often overlooked question to which there are different possible types of responses. For example, a state could compensate anyone who might have been harmed by its exports, or it could take steps to cease performing the harm. An ideal response would include elements of both measures. For reasons of space, I will focus on the issue of ceasing harm as this is the most obvious response to contributing to harm.

Framing a response requires that we be clear about the context. While the Paris Agreement brought about emission reductions targets for states, the enforcement mechanisms to be adopted if states fail to comply or bend the rules are almost non-existent. In the case of exports, the mechanisms by which responsibility for emissions is counted and attributed follow a territorial model which, regrettably, does not allow responsibility for exports to be factored into calculations of responsibility (Eggleston et al., 2006).

Given these factors, what actions ought a state take if it contributes to harm through being complicit? As we have noted, there are strong prima facie reasons not to continue to harm others where we can avoid doing so. Avoiding contributing to harms means not being complicit through producing, supporting and selling fossil fuels as well as influencing decisions in favour of fossil fuels. If that is the case, then the actions a state ought to take should track the kinds of ways it contributes to harm. On the first type of contribution, if a state is to cease to be complicit, it has a prima facie duty not to allow the production and sale of its fossil fuels—restricting supply. Continuing to do so is the most obvious way in which it is complicit in increasing the risk of dangerous climate change.

The second requirement is that states cease *supporting* the production and sale of fossil fuels. The contribution to the increased risk of harms is closely connected to the indirect support that states provide (and facilitate) to the production and sale of fossil fuels, which arises through financial resources such as subsidies and through regulation that favours exports. Supplying infrastructure such as ports or rail lines, tax concessions, cheap finance and favourable regulation may all be crucial to the success of an export industry. Ceasing support of fossil fuel exports is crucial given the role that such support plays in providing benefits to the industry. Focusing just on the sale of exports while continuing to subsidise the fossil fuel industry would still implicate a state in being complicit in harm, especially where the fossil fuel exports were then made available for domestic consumption.

Constraining the finance that is required for fossil fuel exports might also require divestment by states or the entities that they control. Divestment from companies that produce or heavily utilise fossil fuels has become one of the biggest issues in the contemporary moral debate surrounding climate change (Moss, 2017). If investing in fossil fuel export ventures is part of what states or state-owned enterprises do and that investment benefits the exporting process, then these investments ought also to be restricted. By investing in the shares of companies that are 'upstream' producers of fossil fuels, an institution is not directly producing emissions as, say, a coal plant does, or by consuming coal-fired electricity, but it is knowingly investing in part of the causal chain that leads to the emissions. The contribution that an institution makes is thus an indirect one. It does so by providing capital to the relevant fossil fuel companies through share purchases. If this is correct, then the negative reasons to divest stem not from directly causing harm in the sense of creating the emissions, but by knowingly being part of the causal chain that leads to harm, especially where there are alternative courses of action open to the institution.

The kind of regulatory support that is often provided to fossil fuel production is a crucial form of support that deserves comment for two reasons. It is potentially important in allowing production and sale of fossil fuels to continue and is plausibly an appropriate focus of stopping complicity. For example, legislation that allows mining in environmentally sensitive or agriculturally significant areas might be crucial for investors to back a mining venture. The issue of favourable regulation, especially in relation to new mines or production facilities, is also important because some have called for this to be the focus of a response to harms and climate change. In the lead up to the Paris climate talks (COP21) in 2015, civil society groups spearheaded a campaign to 'leave it in the ground' (Lofoten Declaration, 2015). The campaign was an attempt to focus attention on the huge array of potential new coal mines that were mooted in many countries. Given the scale of these new mines, banning new mines was a plausible measure to halt emissions rising further (Voorhar & Myllyvirta, 2013). According to the International Energy Agency and others, in terms of fossil fuel reserves, in order to avoid dangerous climate change, we can only extract and consume one-third or less of the total known global reserves of fossil fuels (Carbon Tracker, 2013; IEA, 2012a, b). Roughly stated, the 'no new coal' proposal was that governments ought not to allow either exploration for new reserves of coal or the granting of new licences to build new coal mines and related infrastructure.

Limiting new mines certainly observes the injunction not to cause harm. By not opening new mines and by limiting the measures that are necessary to do so, states do not risk harming others by allowing new fossil fuel reserves to be produced and used. However, limiting new mines can only be the first step as it does not address the question of what to do about the mines or gas fields that are currently in use and which continue to contribute to the risks of harm.

What to do about limiting the influence of the fossil fuel sector is a more complex question but one that is crucial to the principled governance of a response to climate change (Oreskes & Conway, 2012). It primarily takes two forms. The first is that states ought not to allow fossil fuel interests to have an undue influence on their climate-related policies. This might include restrictions on campaign finance, or allowing recent ex-government ministers to accept paid positions in the fossil fuel industry immediately after leaving office. The second dimension of limiting influence is that states themselves ought not to influence other states or global agreement processes in favour of fossil fuels.

All of these measures focus on limiting the supply of fossil fuels. This is not to say that there cannot be other supply-side mechanisms such as taxing production of fossil fuels (Richter, Mendelevitch, & Jotzo, 2018) or, indeed, demand-side responses such as carbon taxes. However, there are several advantages of supply-side constraints. The first is that they directly target the agent who is contributing to the harm, which is morally preferable to measures that target other agents not involved in the harm. Another advantage is that it is a more emphatic and direct response to the injunction not to cause harm with which we are concerned. If we take the injunction not to harm at face value, ceasing that harm is a more direct and morally clear response than putting in place disincentives such as consumer carbon taxes.

Yet, even if there is a strong prima facie case for imposing supply-side constraints on countries that export fossil fuels, a state such as Australia that benefits substantially from exporting fossil fuels might argue that there are a range of all-things-considered responses that might allow the practice of exporting to continue. For example, if Australia contributed funds to global mitigation efforts, bought offsets or had a fast, low-carbon, domestic climate transition, these actions might balance out the emissions produced by exports. Although these claims cannot be examined here in detail, one response worth mentioning is the claim that exported emissions could be offset. After all, what matters is not the total amount of emissions, but the amount of overall emissions after offsetting or the purchase of permits are taken into account. In taking account of this objection, it should still be noted that while the general point is correct, there is still a prima facie case for states to count at least part of their exported carbon and accept at least some responsibility for the harms it causes. The fact that those harms may be cancelled out by other factors that are part of an all-things-considered judgement does not alter the prima facie claim.

In addition, offsetting measures should not cause harm to others in unacceptable ways, such as conspiring with other governments to remove people from their traditional land or by buying all the cheap emission permits when a state is at its least developed and leaving it little room for its own development-associated emissions. For instance, a less-developed state might initially have to rely on fossil fuels to attain a decent standard of living. Offsetting must also be effective and long term. Offsetting emissions by creating forests in high-bushfire-risk areas or in states subject to uncontrolled land-clearing is clearly undesirable, given that the offset measures risk being destroyed. Second, it should also be noted that purchasing permits or establishing offset programmes are themselves costly and so potentially of considerable impact on the states concerned. These responses aside, where emissions from exports are counted, this adds a strong motivation for a state to take measures to reduce its emissions. This is especially so where either its emissions are already high or its counted emissions from exported fuels are high, or both.

Further, it is noteworthy that the urgency of reduction measures is likely to be heightened by fossil fuel-exporting states having a lower available carbon budget. The point is that there is still a *prima facie* reason to include greenhouse gas emissions from exports in national moral and carbon equations in contrast to current practices. This, then, provides reasons to take responsibility for emissions from exports and the consequences of doing so are likely to be significant.

Some may object that this is too strong a conclusion that the selling and producing of fossil fuels is and has been legal, contributes to a state's prosperity and provides a service (affordable energy) to others that may need to continue for the foreseeable future. Leaving aside some of these all-things-considered objections, there are two responses. The first is that what we are interested in here is the prima facie case for what states ought to do. If there are strong injunctions not to be complicit in harm, then a state ought to heed those injunctions. The second response is that we ought to see stopping in terms of phasing out the production and sale of fossil fuels. This is simply because stopping immediately will not be feasible given that new renewable energy generation infrastructure needs to be put in place. Just how quickly that ought to occur is a complex question that will be a function of the urgency of action on climate change and the likely consequences of the buyers of fossil fuels being able to switch their energy generation to renewable sources.

5   Adjusting the Carbon Budget

Given that the phase out of fossil fuel exports will not be immediate, one potential response to this continuing complicity in harm is including some of the emissions from exports in a state's carbon budget. For example, as part of its 2015 COP21 Paris commitments, Australia has agreed to reduce annual emissions by 26–28% below 2005 levels by 2030. The 2020 target is 532 Mt CO_2-e and the 2030 target is 440–452 Mt CO_2-e (Australian Government, Department of the Environment and Energy, 2015). These targets are likely to be too low to adequately address Australia's fair share (Hueston, Flannery, & Steffen, 2015). Australia's pledges are even less likely to be adequate when contributions to fossil fuel exports are added to this picture. If projected gas export emissions for the year 2020–2021 are added to projected national emissions from direct consumption for 2020–2021, this yields a total of 861.34 MtCO_2-e, which is roughly one-third above the stated target (IPCC, 2014a). While one might not think that Australia ought to bear 100% of the responsibility for the emissions associated with its gas exports, one might nonetheless think that recognising complicity requires that Australia adopt responsibility for *some* proportion of the relevant emissions (such as 50 or 25%). Even though a state is unlikely to be wholly responsible for the emissions that result from its exports, even the allocation of 50% of exported emissions to its domestic budget would likely require significant adjustments to its domestic emission reduction efforts, especially given that many exporting states are already over or at their emissions limit. This result would in turn mean that the transition to a low-carbon economy will need to be far more rapid and costly. Arguably, states such as Australia ought to adjust their carbon budget to take into account the emissions produced by their exports.

6   Conclusion

Whatever the disagreements over whether or not people or states have a duty to aid, causing a harm adds strength to any duty to rectify or compensate for harms in some way. Causing harm in an unjust way should provide a powerful and important constraint on actions (Cripps 2011; Feinberg 1984). It provides *prima facie* reasons for an agent to restrain their actions and/or to be liable for the consequences. The supply, support and sale of fossil fuel exports have now become a significant issue in the race to avert dangerous climate change and fossil fuel exports ought to be accounted for in ways that recognise the moral responsibility of the exporting states. In this way, these states could be made accountable for their complicity in increasing the risks of dangerous climate change. Responding to these risks means that supply-side constraints to these contributions to climate harms should be introduced and implemented. States ought to self-govern their production and sale of fossil fuels even in the absence of enforcement mechanisms to constrain their exports of these harms-causing products.

Acknowledgements I would like to thank Marco Grasso and Alex Lenferna for their comments as well as Beth Edmondson and Stuart Levy for their helpful comments. The research was supported by an Australian Research Council Discovery grant, 'Ethics, Responsibility and the Carbon Budget'.

References

Australian Department of Foreign Affairs and Trade. (2016). Australia to lead Green Climate Fund Board in 2017. Canberra. https://foreignminister.gov. au/releases/Pages/2016/jb_mr_161216.aspx. Accessed September 12, 2017.

Australian Government, Department of the Environment and Energy. (2015). *Australia's emissions projections 2029–2030*. https://www.environment.gov. au/system/files/resources/f4bdfc0e-9a05-4c0b-bb04-e628ba4b12fd/files/ australias-emissions-projections-2014-15.pdf. Accessed October 10, 2017.

Australian Government, Department of Industry, Innovation, and Science. (2016). *Australia's major export commodities: Coal*. https://industry.gov.au/ resource/Mining/AustralianMineralCommodities/Documents/Australias-major-export-commodities-coal-fact-sheet.pdf. Accessed February 2, 2018.

Australian Government, Department of the Treasury. (2017). *Petroleum resource rent tax review*. https://cdn.tspace.gov.au/uploads/sites/72/2017/04/ PRRT.pdf. Accessed October 10, 2017.

Australian Government, Geoscience Australia. (2016). *Gas: Summary*. Retrieved from http://www.ga.gov.au/aera/gas. Accessed November 16, 2017.

Bast, E., Doukas, A., Pickard, S., van der Burg, L., & Whitley, S. (2015). Overseas Development Institute and Oil Change International. *Empty promises: G20 subsidies to oil, gas and coal production*, 32.

Carbon Brief. (2017). Analysis: Just four years left of the 1.5C carbon budget. https://www.carbonbrief.org/analysis-four-years-left-one-point-five-carbon-budget. Accessed January 8, 2017.

Carbon Tracker and the Grantham Institute. (2013). *Unburnable carbon 2013: Wasted capital and stranded assets*. http://www.carbontracker.org/report/ unburnable-carbon-wasted-capital-and-stranded-assets/.

Cassidy, N., & Kosev, M. (2015). Australia and the global LNG market. *RBA Bulletin*. 33–44.

Climate Action Tracker. (2015). Australia set to overshoot its 2030 target by large margin. http://climateactiontracker.org/assets/publications/briefing_ papers/082015_Australia.pdf. Accessed November 2, 2017.

Climate Action Tracker. (2016). Australia. http://climateactiontracker.org/ countries/australia.html. Accessed May 24, 2017.

Climate Council. (2016). *Towards Morocco: Tracking global climate progress since Paris*. Potts Point: Climate Council.

Coady, D., Parry, I., & Sears, L. (2015). *How larger are global energy subsidies?* (Baoping Shang, IMF Working Paper).

Cripps, E. (2011). Climate change, collective harm and legitimate coercion. *Critical Review of International Social and Political Philosophy, 14*(2), 171–193.

CSIRO, Energy Networks Australia. (2016). *Electricity network transformation roadmap: Key concepts report.* http://www.energynetworks.com.au/sites/default/files/key_concepts_report_2016.pdf. Accessed October 10, 2017.

Dediu, D., Ellis, M., Heyning, C., Lambert, P., Segorbe, J. & Thain, A. (2016). Sustaining impact from Australian LNG operations. https://www.mckinsey.com/australia-and-new-zealand/our-insights/sustaining-impact-from-australian-lng-operations. Accessed October 10, 2017.

Eggleston, H. S., Buendia, L., Miwa, K., Ngara, T., & Tanabe, K. (Eds.). (2006). *IPCC guidelines for national greenhouse gas inventories.* Japan: IGES.

Feinberg, J. (1984). *Harm to others: The moral limits of the criminal law.* Oxford: Oxford University Press.

Hueston, G., Flannery, T. & Steffen, W. (2015). *Halfway to Paris: How the world is tracking on climate change.* Potts Point: The Climate Council. https://www.climatecouncil.org.au/uploads/7ec258783dd2367efa806f6dc-6c9a54a.pdf. Accessed October 10, 2017.

International Energy Agency (IEA). (2012a). *World energy outlook 2012.* Accessed November 24, 2017.

International Energy Agency (IEA). (2012b). *World energy outlook 2012: Executive summary.* http://www.iea.org/publications/freepublications/publication/English.pdf. Accessed May 7, 2015.

International Transport Workers Federation. (2016). *Australian LNG Exports to Boom, Tax Revenue is a Bust* (ITF Briefing Paper). https://static1.squarespace.com/static/574507cde707eb332424b26a/t/582d44d3b3db-2bc03f850780/1479361749065/ITF+PRRT+Brief+2+Qatar+-+Sept2016.pdf. Accessed October 10, 2017.

IPCC. (2014a). *Climate change 2013: The physical science basis. Contribution of working group I to the Fourth Assessment Report of the Intergovernmental Panel on Climate Change.* Cambridge and New York, NY: Cambridge University Press.

IPCC. (2014b). *Climate change 2014: Impacts, adaptation, and vulnerability. Part A: Global and sectorial aspects. contribution of working group II to the fifth assessment report of the intergovernmental panel on climate change.* Cambridge: Cambridge University Press.

Kutz, C. (2007). *Complicity: Ethics and law for a collective age.* Cambridge: Cambridge University Press.

Lepora, C., & Goodin, R. E. (2013). *On complicity and compromise.* Oxford: Oxford University Press.

Lofoten Declaration. (2015). *Climate leadership requires a managed decline of fossil fuel production.* http://www.lofotendeclaration.org/#read. Accessed May 31.

Mercator Research Institute on Global Commons and Climate Change. (2017). Remaining CO_2 budget. https://www.mcc-berlin.net/en/research/co2-budget.html. Accessed October 10, 2017.

Moss, J. (2017, October). The morality of divestment. *Law and Policy, 39*(4), 305–428.

Oreskes, N., & Conway, E. M. (2012). *Merchants of doubt.* London: Bloomsbury.

Richter, P., Mendelevitch, R., & Jotzo, F. (2018, February 16). Coal taxes as supply-side climate policy: A rationale for major exporters? *Climate Change.*

Scheffler, S. (1995). Individual responsibility in a global age. *Social Philosophy and Policy, 12*(1), 219–236.

Singer, P. (1972). Famine, affluence and morality. *Philosophy and Public Affairs, 1*(3), 229–243.

Stern, N. (2007). *The economics of climate change: The Stern review.* Cambridge: Cambridge University Press.

UNFCCC. (2015). Report on the structured expert dialogue on the 2013–2015 review. http://unfccc.int/resource/docs/2015/sb/eng/inf01.pdf. Accessed November 2, 2017.

UNFCCC. (2016). Report on the individual review of the inventory submission of Australia submitted in 2015, Note by the expert review team. http://unfccc.int/resource/docs/2016/arr/aus.pdf. Accessed October 10, 2017.

United Nations Office for Disarmament Affairs. (1970). *Treaty on the non-proliferation of Nuclear weapons.*

United States Census Bureau. (2017). *US and world population clock.* https://www.census.gov/popclock/. Accessed October 10, 2017.

Victor, D. G., Akimoto, K., Kaya, Y., Yamaguchi, M., Cullenward, D., & Hepburn, C. (2017). Prove Paris was more than paper promises. *Nature, 548,* 25–27.

Voorhar, R., & Myllyvirta, L. (2013). *Point of no return.* Amsterdam: Greenpeace International.

12

Personal Carbon Trading and Individual Mitigation Accountability

Steve Vanderheiden

1 Introduction

National carbon budgeting forms an essential component of international climate change mitigation efforts. Through it states track the carbon emissions for which they are responsible with a view toward meeting specified decarbonization targets. Under the 2015 Paris Agreement, state parties pledged to follow non-binding national carbon budgets through their Nationally Determined Contributions to international mitigation efforts (NDCs). Through these, in principle, their contributions toward the goal of avoiding 'dangerous anthropogenic interference' (United Nations, 1992) could be assessed. In the absence of legally binding targets or guidelines for what constitutes an equitable contribution, reputational accountability would presumably serve as both motive and enforcement mechanisms for states to adopt ambitious

S. Vanderheiden (✉)
University of Colorado, Boulder, NV, USA
e-mail: steven.vanderheiden@colorado.edu

© The Author(s) 2019
B. Edmondson and S. Levy (eds.), *Transformative Climates and Accountable Governance*, Palgrave Studies in Environmental Transformation, Transition and Accountability, https://doi.org/10.1007/978-3-319-97400-2_12

mitigation targets through their NDCs and to meet those targets by complying with their self-imposed carbon budgets. Concern for national standing and/or pressures from the 'naming and shaming' of unmet or inadequately ambitious NDCs would substitute for stronger accountability measures that were thought to be politically infeasible and thus omitted from the convention.

In the wake of the Paris Agreement, several criticisms of this voluntary and decentralized approach have emerged. Concerns about an ambition gap have arisen from the fact that full compliance with NDCs would fail to ensure the Paris Agreement's temperature target of no more than 2°C of warming. Concerns about compliance with NDCs have also arisen, given the absence of enforcement mechanisms for state pledges, as have those around the equity of such pledges. Nonetheless, the process of filing an NDC and tracking progress toward those state mitigation pledges represents a form of accountability through which the transparency of national carbon budgets is viewed as providing some incentive for states to meet their mitigation responsibilities, and for relevant stakeholders to hold them to account (Keohane, 2006). The 'global stocktake' provides an additional accountability mechanism by which ambition and compliance gaps can be identified and narrowed, and equity in national mitigation contributions encouraged.

To date, the ambition and performance benefits of this monitoring of and reporting on national carbon budgets are limited to national mitigation contributions. No parallel accountability mechanism has been established for holding sub-state actors responsible for doing their part to mitigate climate change. Carbon budgets, through which greenhouse gas emissions are tracked and rationed in a transparent manner, remain underutilized as an accountability tool at the micro-level, where many of the decisions related to national carbon emissions are made and most of the impacts of such decisions felt. This is unfortunate because carbon budgets encourage ambition, reveal inequity and incentivize compliance. States set such budgets through their NDCs and firms can do so through voluntary auditing programs such as the Carbon Disclosure Project, but individual persons or small groups like households currently lack the ability to effectively monitor their carbon footprints and so take responsibility or be held accountable for mitigation actions. The personal carbon budget—that is, the intentional rationing of carbon access among persons rather than states or other large-scale organizations—could help to fill this gap, applying the accountability mechanisms of carbon reporting and rationing across scales.

2   Personal Carbon Budgets and Individual Accountability

While at this stage largely notional, personal carbon budgeting could potentially form a component of global climate change mitigation efforts. Their principles and practices for holding agents responsible for their roles in climate-related harm can be applied across scales, with equity and ambition imperatives for decarbonization applying to persons in a manner that parallels their application to states. Indeed, the transition to a low-carbon society requires attention to the allocation of carbon access among persons as well as between states if its equity imperatives are to be met, even if such imperatives have thus far largely been developed for and applied to international rather than interpersonal inequality. Currently, remedial responsibility for climate change is differentially assigned to states under legal and ethical principles for informing the design of international mitigation efforts. It could also be applied with respect to persons, on the basis of many (if not all) of the normative criteria that climate justice scholars have developed for states. This chapter explores the normative criteria and develops some relevant principles for considering the personal carbon budget as a potential accountability tool as well as a conceptual expression of interpersonal equity in mitigation responsibilities.

Indeed, for states to implement the mitigation targets that they have been assigned under national carbon budgets, they must pass along those targets in what David Miller (2008) calls a 'two-stage approach' by which shares of national abatement obligations are allocated among sub-state actors, including persons. According to Miller (2008, p. 121), states 'may decide to control emissions by taxing the industries that mainly produce them, or they may decide to give each individual citizen a carbon budget that limits their use of emission-generating resources to a total that they can exceed only by buying a slice of somebody else's', and they do so 'according to guidelines that are agreed internally'. Hence, states cannot avoid confronting many of the same normative issues in implementing national mitigation targets as have been evident in efforts to set those targets within an international climate treaty framework, nor can they avoid assigning differentiated remedial responsibility among citizens, even if they refuse to acknowledge doing either.

Most analysts view a carbon tax and some kind of carbon rationing system as the two main policy alternatives. In deciding upon implementation measures, states should seek to design domestic carbon abatement programs and strategies that are capable of differentiating burdens among persons and other sub-state actors in accordance with defensible normative criteria. While a carbon tax would be less costly to implement, a personal carbon budgeting scheme offers several key noneconomic benefits that are unavailable without the carbon visibility and personal carbon entitlements found in rationing schemes that allow for trading of unused emissions permits (or 'cap and trade' schemes). Personal carbon trading (or PCT) schemes are able to more closely approximate climate justice demands for assigning responsibility for climate change. They do so through a form of personal carbon budgeting that mobilizes a normative sense of individual responsibility that both justifies and motivates domestic mitigation efforts. While firms and other collective sub-state entities might also be assigned emissions caps in implementing national mitigation targets, the focus here is upon PCT schemes for implementing those targets and assigning individual remedial responsibility for climate change.

In order to focus upon several core issues in individual carbon rationing, I shall bracket several problems related to assigning individual emissions caps or allowing their trade through an offset market. First, I assume a defensible series of annual global emissions budgets capable of satisfying climate justice objectives, and that these can be justly allocated among the world's states. Second, I assume that national annual emissions budgets can in principle be justly allocated to sub-state parties, including resident persons. My primary aim here is not to explore the resource-sharing principles by which particular shares of remedial responsibility for climate change might be calculated, but rather to consider instruments through which this responsibility could be discharged. Finally, I assume that compliance with national and personal caps could be effectively monitored and enforced, through a transparent system in which parties are aware of their current and past emissions as well as those associated with their future options. All three assumptions bracket serious problems associated with PCT schemes that must be addressed before any such scheme is tenable in practice, but are not of direct interest to this inquiry.

The first set of issues concerns the mobilization of personal responsibility through carbon budgeting under a PCT scheme, which despite its costs and implementation difficulties is viewed as practically feasible if applied to limited carbon sources such as transport and energy use. As Tina Fawcett notes (2010, p. 6857), in the UK PCT is 'considered to be more acceptable than the alternatives of direct or indirect taxation' although it is 'not yet a fully worked-out policy'. Most practical PCT proposals limit their purview to emissions from transport and household energy use because it is relatively easily monitored at the pump and through monthly utility accounts. These proposals require persons to pay from two distinct accounts in purchasing carbon-intensive goods and services: first to pay the market price for the good or service, and then to deduct its carbon credits from their personal allowance. Various proposals include core features such as the allocation of individual carbon allowances to cover their own emissions, which is periodically replaced and which declines annually, along with the national emissions budgets from which they are derived. Most versions withhold a share of the national emissions budget from allocated personal carbon allowances, to account for emissions from firms and other sub-state actors and also supply additional carbon credits for purchase by individuals that exceed their individual quotas. Prices for these extra carbon shares reflect market supply and demand, fluctuating over time in a manner that carbon taxes do not.

In order to focus upon the power of personal carbon budgeting in comparison with carbon pricing mechanisms that do not entail personal CO_2 emissions budgets, a PCT scheme, through which persons hold carbon permits as a tradable commodity is compared with a carbon tax, through which carbon is taxed without rationing or trading systems. Given its potential to activate and mobilize a sense of remedial responsibility, which turns on what I call *cognitive responsibility*, several key advantages to PCT appear to obtain, including the more widely observed efficiency and autonomy benefits of personal trading. An additional benefit follows from the system's instantiation of equity norms for personal carbon consumption and provision of carbon budgeting feedback on various consumption choices. With PCT seemingly able to deliver these noneconomic benefits over a carbon tax, several advantages over carbon trading schemes at other levels can be identified. These include upstream rationing among firms or to states; considering three objections that have been lodged against carbon trading; and, finding PCT to be considerably less vulnerable to such critique than upstream trading schemes, analysis finds that PCT potentially offers a system capable of realizing the benefits of carbon trading without incurring its biggest flaws.

3   Taking Responsibility: Is Paying Enough?

Practically speaking, a well-designed carbon tax can be an effective instrument for reducing emissions and financing further decarbonization efforts. Both mechanisms create conservation incentives in rationing and/or pricing carbon, and both would need to be designed to account for their allocation of burdens among parties in order to satisfy justice imperatives. Many economists favor it to an emissions trading scheme (ETS) due to its relative ease of implementation and for the regular revenue stream that it can yield. Since it prices all carbon rather than setting aside personal or group allowances, a carbon tax sidesteps the controversy surrounding the allocation of shares to various parties, which an ETS must address. However, a tax also has several disadvantages, compared to an ETS. Since it lacks a hard cap on allowable emissions, a carbon tax offers no assurance that states will adhere to national carbon budgets. Instead, it relies upon elastic demand for carbon from which incentives to reduce consumption or seek substitutes are created. Like an upstream rationing scheme, which passes along carbon pricing to consumers through higher energy and transport costs, a carbon tax can be regressive if basic access to energy is not subsidized or low-income energy users are not compensated. The most significant differences however concern the relative invisibility of a carbon tax and its absence of an individual emissions entitlement, as shall be discussed further below.

Domestic climate change mitigation efforts through which persons are merely made to pay for the carbon they use must be further distinguished from ones in which they are placed on a carbon budget and informed about how their various alternatives affect compliance with it. These differ in the ways that each holds agents responsible for some harm toward which they contribute. An agent takes *remedial* responsibility by acting to mitigate or avoid some harmful outcome that would otherwise occur, or to rectify some harm that has already occurred. In his influential account, Miller (2001, p. 454) describes such responsibility as involving criteria by which agents are 'picked out, either individually or along with others, as having a responsibility towards the deprived or suffering party that is not shared equally among all agents'.

Under an international climate treaty framework, states would be assigned remedial liability in accordance with their 'common but differentiated responsibilities and respective capabilities' and would implement these corrective justice obligations through the subsequent assignment of remedial responsibilities among sub-state parties within their borders. As Miller suggests, this responsibility is owed to 'the deprived or suffering party' and not merely to some state tax office. However, remedial responsibility can sometimes be discharged by paying into schemes through which remedies to harm are made available. Insofar as a carbon tax raises revenues that could be directed toward domestic or international decarbonization efforts that reduce the causes of climate change or toward adaptation efforts that seek to reduce the human suffering associated with its effects, while also creating economic incentives to reduce carbon emissions, it could constitute such a remedy. Persons paying that tax—whether or not they are aware that they are doing so, or that its proceeds are being used to provide a remedy to problems caused by the pollutant to which the tax is attached—could be viewed as exercising *economic responsibility* (or being held economically responsible) for the harm in question. This form of responsibility requires no admission of fault or even recognition of the harm or victims toward which proceeds are directed, as it merely involves the bearing of remedial costs, and so constitutes one variety of remedial responsibility.

In his account, Miller identifies an agent's moral responsibility for faulty contributions to the harm in question as the strongest criterion for assigning remedial responsibility. He also notes that other criteria sometimes apply where moral responsibility cannot be attributed, including mere causal responsibility, such as when contributory actions cannot be faulted. Other criteria include the capacity to assist, and special ties of community with victims. In each instance, he argues, the 'overriding interest' in assigning remedial responsibility is to 'identify an agent who can remedy the deprivation or suffering that concerns us' (Miller, 2001, p. 471). Where imperilled victims require immediate attention if the most serious harm is to be averted, remedial responsibility is urgently needed. In these cases, the expeditious assignment of responsibility can involve one agent being assigned to act as a first responder and another later being required to finance that initial action or compensate the first agent for any burdens incurred. This approach pertains when proximate and capable agents undertake an expensive rescue but are then compensated for the costs of doing so by morally responsible parties. A carbon tax could be viewed as a parallel instance involving a form of economic responsibility through which persons help to finance more immediate remedies, along with the nudge toward remedial decarbonization that the pricing mechanism also provides.

Another form of responsibility, which is not directly remedial but which can assist in supporting remedial actions as well as reducing the need for them, often manifests alongside one or more responsibility-expressing mental states. With or without taking economic responsibility for some harm by contributing toward the effort to mitigate it, an agent can take *cognitive responsibility* through the conscious recognition of the harm in question along with the agent's role in it, with the acknowledgment of an obligation to respond appropriately. Since remedial responsibility may be assigned to persons other than those contributing toward some harm, as with Samaritan duties to rescue based on proximity to the victim and capacity to assist, cognitive responsibility requires neither causality nor fault, but acknowledges some basis for taking on some remedy. Neither must it require capacity to provide an adequate remedy. Agents could take cognitive responsibility for some harm that they are powerless to prevent, discharging that responsibility either through vicariously remedial actions that seek to protect others from similar harm in the future or through mental states like agent regret or atonement that express this responsibility to oneself or others, and which seek to offer a non-remedial response to it. All such responses are additional to cognitive responsibility, through which the agent takes account of their role in some harm, if not as responsible for causing it then as having some obligation to respond to it in some way. This cognitive aspect of remedial responsibility is distinct from the remedial action itself and is often viewed as the essence of *taking* responsibility, which entails recognition and acknowledgment in addition to some kind of action.

The distinctive contribution of PCT compared with upstream rationing measures or a carbon tax is then that it encourages persons to take cognitive responsibility for their role in their country's carbon footprint (if not climate-related harm itself), which the PCT scheme identifies as harmful by dint of the personal limits it prescribes. Andrew Dobson (2006) describes the cognitive responsibility for global environmental harm issuing from awareness of ecological space use patterns as generating a 'thick cosmopolitanism'. He argues that there are motivational advantages in remedial actions through implied chains of cause and effect that are not fully considered in accounts of global ethics based in humanitarianism or distributive justice. Advocates of informational governance likewise tout such benefits. They claim that disclosure and transparency requirements can motivate pollution avoidance through a combination of the empowering effects of informational feedback about alternative actions that agents are considering undertaking and the reputational accountability that disclosure of environmental performance data provides.

A PCT system promotes cognitive responsibility by instantiating a norm of equitable access to carbon sinks and by providing persons with regular feedback concerning the impact of various actions upon their personal carbon footprints. It identifies an individual carbon entitlement through the rationing scheme, beyond which persons face sanctions (in the form of a fee for additional shares) for excessive emissions. In addition to ensuring economic responsibility for personal emissions that exceed per capita entitlements through the requirement of purchasing additional carbon credits, PCT schemes reinforce the norm through this sanction for exceeding one's equitable share. They also provide additional feedback through the increasing or decreasing per-unit cost for additional carbon credits, which reflect overall social demand for such additional credits and thus social progress toward decarbonization imperatives. They raise consciousness about the drivers of high-carbon consumption patterns and the availability of low-carbon alternatives. Promoting cognitive responsibility for each person's contribution toward national mitigation targets could assist in overcoming norms that enable unsustainable lifestyles and transforming attitudes and beliefs surrounding greenhouse pollution and climate change.

4   Carbon Trading and Its Discontents

The cognitive aspect of responsibility-taking through a personal carbon rationing scheme does not require that carbon trading be allowed. Individuals would still need to be made aware of their carbon footprints along with the carbon content of their consumption choices, but there may be solidaristic benefits of PCT schemes that depend upon market signals from a trading system. As David Fleming (2007, p. 14) notes of tradable emissions quotas, trading could foster a sense of common purpose from which a more cooperative ethos for developing a sustainable society may emerge, rather than individualizing and depoliticizing decarbonization efforts through taxes or rationing schemes that prohibit trading.

First, the fixed quantity makes it obvious that high consumption by one person leaves less for everyone else. Your carbon consumption—that is, the extent to which you depend on fossil fuels—becomes my business: I have an incentive to influence your behavior to our mutual advantage: Lower demand means lower prices… Secondly, the big structural changes—including a substantial localization of the energy system—that will be needed to achieve deep reductions in dependency on fossil fuels will not by any means be simply a function of individual effort.

This proposal involves trading options within an individual rationing scheme, in order to implement national mitigation obligations and to assign remedial responsibility among sub-state parties according to justified criteria. Despite the above-noted benefits of carbon trading, three objections to it are examined below.

The first objection concerns the commodification of either CO_2 itself or its sequestration capacity, which allegedly results in the inequitable allocation of goods or services from linking unequally held economic power with the permission to emit greenhouse gases. Another objection concerns the commodification of ecological goods and services without reference to any further impacts of their market allocation, claiming commodification of nature to be wrong in itself. Such an objection would apply to all carbon pricing systems rather than cap and trade schemes like PCT. Trading systems inherently exacerbate or exploit existing wide inequalities among and between people by associating one kind of disadvantage-conferring inequality with another. This claim is thus that carbon trading is *unjust*. The second objects to the delegation of abatement obligations to others through trading on consequentialist grounds. It claims that delegating decarbonization obligations to others through trading rather than undertaking them by oneself slows the transition to a low-carbon society and economy, where urgency requires that transition to be made quickly. Here, the focus is upon impacts other than those associated with socioeconomic inequality, and the claim is that carbon trading is *bad*. Finally, the third objects to delegating such obligations for reasons related to assignments of moral responsibility. This claim is that carbon trading is *irresponsible* because offset purchasers are evading some abatement obligations that are properly their own.

The standard economic case on behalf of carbon trading is consequentialist, arguing that trading allows for more efficient decarbonization than would be possible without it, and may be more politically feasible in states with commitments to neoliberal market trading regimes for other goods. Because carbon trading allows agents to utilize the decarbonization options with the lowest per-unit abatement costs, rather than requiring that they reduce their own emissions at potentially much higher per-unit cost, they can reach their abatement targets more cheaply. The relative economy of allowing for more efficient abatement options would not directly benefit climate change mitigation efforts. However, compared against parties reaching their targets through more expensive direct abatement efforts, the lower costs associated with a trading system might make the ETS under which trading is conducted more politically feasible, or might allow its decarbonization goals to be more ambitious. One might argue from opportunity costs that spending more on direct decarbonization would lead to morally worse outcomes since it might divert resources away from other important efforts like poverty relief, but efficiency alone would not otherwise commend such schemes on ethical grounds.

Indeed, John Broome (2012, p. 93) makes such an argument, defending the use of carbon offsets for their efficiency and claiming that 'as a general rule, it is better for the world if things are done where they can be done most cheaply'. Arguing for a duty to achieve carbon neutrality on justice grounds rather than by appeal to consequences, he acknowledges that persons could do more good by using their money on poverty relief or public health efforts rather than purchasing carbon offsets. In his view, maximizing goodness would involve 'acting unjustly by emitting greenhouse gas that harms people' (Broome, 2012, p. 91). Distinguishing between duties of beneficence or humanity that oblige persons to aid the vulnerable and duties of justice that oblige them to avoid causing harm, Broome argues that the latter have priority and are more stringent. At best, more efficient means of fulfilling one's duties of justice by achieving carbon neutrality would be instrumentally and contingently good, provided that the resulting savings were invested in humanitarian efforts to improve the lives of others, not used to enhance one's own consumption opportunities. Here, efficiency makes ethical action possible, but has no moral content of its own. Notably, Broome objects to the compliance offsets used to meet national decarbonization targets, referencing only the voluntary offset market through which personal emissions are not capped and therefore cannot be traded. His qualified and indirect defense of efficiency might likewise however apply to a PCT scheme, given its potential to allow agents to do further good beyond the demands of justice.

Simon Caney (2010) likewise points to considerations of efficiency and feasibility in defending carbon trading schemes, finding these to be pragmatically justified at the country or firm level if also objectionable at the personal level. Having noted that economic instruments like carbon taxes or trading schemes allow parties to discharge a given abatement obligation at the lowest cost, he identifies opportunity costs of undertaking more costly abatement options. Caney and Hepburn (2011, p. 206) suggest, along with Broome, that 'these wasted funds might have been used to develop new low-carbon technologies and products, increased staff wages, been passed onto shareholders or simply given to charity'. Elsewhere, Caney (2010) concedes that international emissions trading might not actually lower emissions, as the cap rather than the various compliance options are what ensures results. He suggests (Caney, 2010, p. 216) that if trading is 'a persuasive sweetener' to reluctant parties and if 'powerful actors sign up to the package as a whole only because' of it, then it would be 'wrongheaded' to reject trading 'even though it does not itself lower any emissions'.

The efficiency claims touted by Broome and Caney depend upon two premises that are contested by critics. The first concerns this contingent value, where resources saved as the result of more efficient carbon abatement lead to greater national support for humanitarian causes or redoubled sustainability efforts, for which experience suggests a justified skepticism. The second depends upon the claim, accepted by both Broome and Caney but doubted by scholars of various carbon offset programs, that 'emitting a tonne of carbon dioxide and offsetting it is exactly as good as not emitting it in the first place, providing the offset is genuine' (Broome, 2012, p. 89). Offsets must be equivalent in their physical effects to be counted as such, which Broome doubts in the context of sequestration offsets developed under the Reducing Emissions from Deforestation and Forest Degradation in Developing Countries framework (REDD). States are expected to trade internationally in order to meet their mitigation targets under this successor treaty to the expired Kyoto Protocol. Against the 'genuine' quality of offsets from reforestation projects, Broome (2012, p. 95) claims that the use of offsets by states in meeting their national mitigation targets 'will simply lead to extra global emissions unless any new carbon credits it produces are balanced by a corresponding cut in emissions permits around the world'. He argues that when used to achieve compliance with national caps, offsets offer developed countries 'a useful smokescreen for evading their responsibilities' (Broome, 2012, p. 94). Curiously, the voluntary carbon offsets that persons use to achieve carbon neutrality evidently enjoy this moral equivalence for Broome, since he endorses schemes by which paying someone else to take remedial action is fully equivalent to undertaking that action oneself. This is despite the far less rigorous standards for ensuring biophysical equivalence among the voluntary offsets that he commends compared to the compliance offsets that he rejects.

By contrast, Caney (2010) implicitly endorses the use of compliance offsets in international carbon trading schemes and in trading among firms present in upstream rationing systems. He rejects personal carbon trading, citing an argument about autonomy expressed in individual terms on behalf of trading by states or firms. He approvingly cites Simmel in claiming that 'allowing persons to discharge their duties through the payment of money rather than through performing specific acts or providing in-kind payments grants them a greater degree of freedom than they would otherwise have' (Caney, 2010, p. 200), despite rejecting such trading when done at the individual rather than international level. Nonetheless, one might defend PCT schemes on the basis of the autonomy that they allow, especially in the greater autonomy afforded to individual persons.

States that opt to sell carbon credits that would otherwise allow their citizens to experience the benefits of development do not grant those citizens a 'greater degree of freedom', especially when they fail to consult or override the expressed preferences of many of those citizens. Those draining national treasuries to purchase additional carbon credits rather than using those funds to develop the low-carbon infrastructure that many of their citizens prefer likewise fall short of the autonomy value that Caney finds in international carbon trading. If freedom or autonomy is to serve as a reason for endorsing carbon trading, it would seem most compellingly located in a PCT scheme by which persons make decisions that affect themselves alone, not in international or upstream schemes by which a few elites make trades that affect persons whose consent is not required or solicited in advance. However, the efficiency and autonomy benefits of carbon trading schemes are often opposed by objections thought by some to outweigh them, and to these we now turn.

4.1   Objection 1: Equity

By pricing the permission to emit CO_2 and allowing it to be bought and sold, PCT commodifies either the pollutant or the sink capacity that allows it to be harmlessly emitted. With this commodification, critics argue, ecosystem services like sink capacity 'become the basis for new socio-economic hierarchies, characterized by the re-positioning of existing social actors, the emergence of others and, very likely, the reproduction of unequal power relations in access to wealth and environmental resources' (Kosoy & Corbera, 2010, p. 1234). Given the correlation between carbon emissions, energy use, and privileges associated with affluence such as greater mobility or higher consumption rates, the commodification of carbon allowed through trading schemes has the effect of distributing such privileges by market supply and demand. This has the effect of allowing existing economic inequities to translate into inequities in access to ecosystem services or the privileges that these allow.

David Harvey (2007, p. 35), for example, claims that the 'primary aim' of commodification 'has been to open up new fields for capital accumulation in domains formerly regarded off-limits to the calculus of profitability', leading to 'accumulation by dispossession' through which resources transferring from common to private ownership shift wealth from rich to poor. In the context of international carbon trading, Bumpus and Liverman (2008, p. 144) consider this form of 'accumulation by decarbonization', an unequal exchange between rich and poor that 'disadvantages others who were more efficient or less powerful in negotiations or were willing to assign their carbon rights to others at a low cost - such as forest owners in the developing world'. By allowing the world's affluent to buy cheap carbon credits on offer from within developing states, at rates that reflect the latter's disadvantaged bargaining position, trading based on the creation of carbon credits through sequestration projects in the developing world can allow the affluent to profit by exploiting cheap offset opportunities in the global South.

Such a critique applies primarily to the compliance offset system through which carbon credits are traded among states or other large organizations, and where some credits are created through projects undertaken in poor states. Since PCT allows trading among only domestic parties and prohibits the sort of 'carbon colonialism' referenced, the kind of exploitation that critics attribute to international carbon trading would not apply. While economic hardship may lead some to sell personal carbon credits from what resembles a kind of coercion in PCT schemes, buyers acquire only the one-time permission to emit rather than the sink capacity that makes ongoing emissions possible, as with the international carbon trading systems referenced above. Since those exceeding their personal carbon allowance would buy extra credits on an exchange rather than purchasing them directly from sellers, they could not exploit unequal bargaining positions in the way that they might under so-called carbon colonialism systems. At least with regard to this form of the equity objection to carbon trading, PCT schemes appear to be less vulnerable than the more commonly endorsed international trading schemes.

Another equity worry may apply at the domestic level and through the trading of carbon among persons. As Jonathan Aldred (2012, p. 343) notes, 'carbon trading extends the domain of distribution of goods based on willingness to pay (and ability to pay) in the market', allowing the affluent greater access to those goods and activities with embedded carbon. Given existing economic inequality among persons, market trading would allow 'extreme inequality of access' that currently characterizes market-distributed goods to be 'spread to more goods', including those related to energy use. Unless corrected in some way, Aldred claims, the effects would be highly regressive. Since CO_2 'is a prerequisite for the fulfillment of basic needs', its distribution by market principles can be 'akin to regressive taxation' in that 'the burden of a higher carbon price falls more heavily on the poor, because they spend a higher proportion of their income on goods whose production requires carbon emissions' (Aldred, 2012, p. 345).

Aldred's concern about regressivity in the impact of carbon pricing applies more directly to carbon taxes or upstream rationing schemes involving firms, which pass costs on to consumers through higher prices, than it does to a downstream scheme like PCT, which grants all persons some basic emissions entitlement that allows for the protection of basic needs. Similar to carbon tax compensation schemes that provide some basic access to carbon before the tax applies, or compensate low-income persons for expenses related to basic needs, a PCT guarantees a basic carbon entitlement that should blunt the worst aspects of the individual equity critique. Indeed, a PCT is unique among carbon pricing schemes in that it guarantees free access to that carbon that is necessary for meeting basic needs along with some level of luxury emissions, even if it would allocate further luxury emissions according to ability to pay. As Spiekermann (2014) notes, surplus carbon credits would be distributed by market supply and demand under a PCT, potentially allowing the affluent greater access to goods like airline travel that require the purchase of credits beyond this personal entitlement, but the PCT itself would narrow rather than widen existing inequalities related to access to ecosystem services. A PCT cannot bear all the fault for being unable to neutralize the insidious impact of socioeconomic injustice. This is reasonably the case even when it aims to more equitably allocate access to goods that would be disproportionately controlled by the rich without such a scheme with bad environmental consequences of climate change falling largely upon the poor. Indeed, as noted above, PCT schemes appear to better promote equity in carbon access than do any other carbon pricing mechanism available for use in implementing national mitigation targets, obviating the force of this objection to trading.

4.2   Objection 2: Consequences

A different objection arising from the commodification of carbon pollution or associated sink capacity concerns the expressive content of the permission that market trading implies. In a frequently cited version of this critique, Michael Sandel (2005, p. 94) argues that 'turning pollution into a commodity to be bought and sold removes the moral stigma that is properly associated with it'. Whereas an *ex post* fine for polluting preserves that stigma by 'the community conveying its judgment that the polluter has done something wrong', Sandel (2005, p. 94) explains, an *ex ante* fee conveys a permission to engage in harmful activity.

The distinction between a fine and a fee for despoiling the environment is not one we should give up too easily. Suppose there were a $100 fine for throwing a beer can into the Grand Canyon, and a wealthy hiker decided to pay $100 for the convenience. Would there be nothing wrong in his treating the fine as if it were simply an expensive dumping charge?

By implying this implicit permission to pollute, he argues, marketable pollution rights (of which carbon offsets offer an imperfect example, to be discussed below) allow persons to act against community goals like environmental protection so long as they pay enough to do so, which 'may undermine the sense of shared responsibility that increased global cooperation requires' (Sandel, 2005, p. 95).

Sandel's critique might be viewed as indirectly consequentialist; as claiming that fines are superior to fees because of their expressive value, which reinforces social norms against pollution and so further discourages it. Here, one might distinguish between the intrinsic motive of reducing one's pollution on the belief that it is wrong and the extrinsic motive of reducing it because of some economic cost attached to it. The claim might thus be restated: While both fines and fees attach an economic cost to polluting and thereby furnish an (equal) extrinsic motive for avoiding it when possible, fines reinforce the social stigma attached to violating a norm. They thereby furnish an additional intrinsic motive for doing the same, while fees connote permission and thus undermine that intrinsic motive. Understood in this way, Sandel's critique does not endorse the claim that pollution rights would be inappropriate to exchange for money or to allocate through a market, with which it is sometimes conflated, since both fines and fees in effect allow persons to pollute in exchange for a charge that is identical in both cases. His critique therefore also applies in one sense to any carbon pricing mechanism, all of which allow persons to emit some amount of CO_2 in exchange for a fee, urging instead a regulatory scheme whereby some level of personal carbon pollution is prohibited, with violations punishable by a fine. Note, however, that only a PCT scheme expresses a norm of equitable emissions through which one might infer some social disapprobation against excessive pollution, as Sandel recommends, rather than treating all pollution as fully permissible so long as required fees are paid.

Under upstream rationing schemes whereby firms are required to purchase carbon credits (or pay a carbon tax), a similar effect might occur. As Aldred (2012) suggests, where firms are allowed the option of paying a fee to pollute (such as through offset schemes) rather than being fined and implicitly chastised for polluting in excess of permitted levels (such as through 'command and control' regulation), corporate social responsibility (CSR) programs designed around carbon emission limits may be undermined. In such cases, the fees would allow firms the option of achieving their abatement targets through offsets rather than by reducing their own emissions. Echoing Sandel's worry about expressing a moral permission to pollute so long as 'the requisite number of permits are purchased', Aldred (2012, p. 351) claims that 'carbon trading weakens the stigma attached to a large carbon footprint and, therefore, the reputational gain to be had from reducing it'. Shareholder movements to pressure firms into taking on carbon reduction goals may be frustrated insofar as those firms can receive the same reputational benefit by simply purchasing carbon offsets rather than changing behavior or upgrading infrastructure, forestalling meaningful progress toward low-carbon modernization. A similar dynamic in deciding between the purchase of extra carbon credits and undertaking personal carbon abatement efforts may plague PCT schemes. Although some might be allowed to shirk their share of social decarbonization efforts by paying others to do this for them, the full biophysical equivalence of both alternatives makes PCT less vulnerable to this consequentialist objection than upstream rationing schemes. Indeed, to return to Sandel's analogy, the wealthy hiker would be paying someone else to pack the beer can out of the canyon rather than paying a fee to pollute the national park, which might make the activity vulnerable to the third objection but not to one predicting increased levels of pollution.

Other consequentialist objections against the commodification of the ecosystem service related to carbon sequestration are available, however, as the process of commodifying carbon can lead to undesirable outcomes through its incentive effects. Rendering carbon sequestration capacity as a tradable good through schemes like REDD requires the *individuation* of this ecosystem service. This involves imposing legal boundaries around certain phenomena so that they can be bought and sold (Castree, 2003), as well as *itemization,* which 'results from the separation of such biological function from existing forests or from future planted trees or forested areas', and measuring that capacity 'through biomass content and growth models which translate it into tons of carbon dioxide stored in trees' (Kosoy & Corbera, 2010, p. 1231). In practice, this has led to 'the conservation and planting of certain tree species above others, such as those with the largest carbon content or higher growth rates' and has encouraged states and private landowners 'to invest … in tree plantations more than encouraging the restoration or conservation of complex tropical or subtropical ecosystems', thereby harming biodiversity and ecosystem resilience (Kosoy & Corbera, 2010, p. 1231).

Note again that such consequentialist objections against carbon trading apply primarily to schemes other than a domestic PCT, which commodifies carbon emissions without allowing the buying or selling of carbon sinks, and which therefore yield none of the harmful incentive effects noted above. Worries about effects upon forestry practices apply to international schemes such as REDD that allow credits to be created through reforestation projects, and Aldred's objection applies to carbon trading among firms. Only Sandel's critique of the lost expressive value of prohibitions would apply to the kind of commodification that PCT schemes entail. Personal attention to decarbonization imperatives can help to instantiate new low-carbon norms through the expression of disapprobation for the *excessive* and harmful personal emissions. There is normative potential from introducing new restrictions on carbon emissions where none existed previously. Only PCT can promise the efficiency and autonomy associated with carbon trading without the potential for negative consequences associated with either upstream firm-based or international trading.

4.3   Objection 3: Responsibility

A more cogent formulation of Sandel's critique of carbon trading might be construed in terms of responsibility rather than consequences. Here, the claim is that allowing pollution in exchange for a fee fails to hold agents responsible for doing their part to reduce society's overall levels of pollution. The wealthy hiker would be shirking her duties to reduce littering within the canyon, which require all users to bear certain convenience costs, rather than allowing those with the financial means to delegate those duties to others. As noted above, this would not necessarily result in more pollution, as Sandel implies, but it would shift the burden of undertaking required abatement activities from the buyer to the seller of personal carbon credits. The payment in exchange for the transfer of carbon credits grants legitimacy to this delegation, according to this view, rather requiring pollution abatement duties to be the responsibility of each.

Internationally, this objection to the delegation of decarbonization activities by developed to developing states motivated several restrictions upon compliance offsets in international carbon trading under the Marrakesh Accords, adopted at the 2001 Conference of the Parties (COP-7) to the United Nations Framework Convention on Climate Change (UNFCCC). It declared that developed states should 'implement domestic action in accordance with national circumstances and with a view to reducing emissions in a manner conducive to narrowing per capita differences' between states. Thus, the use of market-based compliance mechanisms such as emissions trading 'shall be supplemental to domestic action and that domestic action shall thus constitute a significant element of the effort made by each Party included in Annex I to meet its quantified emission limitation and reduction commitments' (UNFCCC, 2001a). At the international level, the ability to delegate carbon abatement duties through offsets and international trading was viewed as a form of wrongful shirking, through which those parties most responsible for climate change were seen as failing to do their part in combatting it.

Caney and Hepburn (2011, p. 214) term this the *Collective Sacrifice Argument*, holding responsibility for climate change to involve 'non-delegable duties' in which agents are required to 'constrain their own emissions and not pay for someone else to lower their emissions'. They reject it as appealing to autonomy in allowing for exchanges between buyers and sellers of permits and question why decarbonization efforts should invoke a sense of civic responsibility in the first place. Since their interest lies in emission trading among firms or states rather than individuals, Caney and Hepburn's primary concerns lie with efficiency, to which they appeal in asking whether it would be permissible for a rich Western state to pay a developing state like China for compliance offsets rather than reducing its domestic emissions. Here, they presume that none could object to a rich state receiving credit for paying a poor state to lower its emissions when this could be done more cheaply than achieving the same level of carbon abatement at home, considering only the scenario where the rich state 'can't really be bothered' to take domestic action and so purchases the carbon credits from a poor one to delegate that action for noneconomic reasons. Here, they note an equity objection, supposing that 'it might seem problematic for the wealthy to pay the poor to forego a good that the wealthy continue to enjoy', but dismiss the objection on grounds that its remedy requires that there be a 'fair distribution of resources, including a fair share of emissions permits' globally (Caney & Hepburn, 2011, p. 217). A similar remark might be made on behalf of a PCT, although Caney and Hepburn are concerned here not with a personal trading scheme but with one among states.

Elsewhere, Caney (2010, p. 208) concedes that an objection about the delegation of duties through an offset scheme is 'most plausible' against PCT schemes, which involve a cooperative scheme 'in which everyone is charged to perform a specific shared task as a civic responsibility', and where trading would allow some to exempt themselves from 'acting in a public-spirited way furthering the common good'. However, he claims that in carbon trading 'the motivation of the person lowering emissions does not matter at all' for the outcome (Caney, 2010, p. 209), and rejects the notion that persons must reduce their emissions for intrinsic moral reasons related to their doing their part in a collective scheme rather than from the economic motivation associated with the offset scheme, dismissing civic responsibility-based objections to trading altogether. Likewise, Page (2011, p. 267) argues that such an objection 'fetishizes the importance of agential responsibility' while failing to consider that 'this seems to be a price worth paying if the result is a more efficient response to climate change', thus undercutting an objection that he finds 'troubling' but 'hardly decisive' (Page, 2011, p. 268).

It is thus fitting that collective sacrifice forms a core component for the case on behalf of PCT schemes, which allow trading but nonetheless call for a kind of collective sacrifice in that all must comply with rationing schemes through which all are affected. In advocacy of PCAs, for example, Keith Hyams (2009) reiterates the importance of acting from a sense of common purpose. He notes that the economic motive for trading under a PCT scheme 'would be supplemented by the additional moral motivation accompanying the belief that one is contributing one's fair share to the burden of discharging a collective responsibility' (Hyams, 2009, p. 238). Critical to this motive is the cognitive responsibility that accompanies taking economic responsibility through a combination of personal decarbonization and offset trading, which Hyams argues would not be diminished by providing for trading unused credits among persons. Critics directing their ire against carbon trading for putatively allowing for personal responsibility to be delegated would do well to consider how it can also help to found such responsibility. They should also consider how limits to such delegation under well-designed PCT schemes might be able to minimize the force of this objection. Surplus emissions credits would be unaffordable for anyone if nobody undertook significant carbon abatement actions under a PCT designed to reduce domestic emissions to the extent required by just national mitigation targets (Sagoff, 2002). Limits on selling personal carbon credits beyond the subsistence threshold similar to those proposed for states under the Marrakesh Accord (UNFCCC, 2001b) could also further ensure that all bear some abatement burden in reducing their society's emissions rather than fully delegating these to others.

Collective sacrifice, like the 'common but differentiated responsibilities' principle from the UNFCCC, requires that all be held responsible for climate-related harm in some way, and in accordance with ethical criteria by which remedial responsibility is properly assigned. As with international schemes, a well-designed PCT will set defensible limits upon trading to protect basic needs and to ensure that all do their part. This would address the worries that critics express about any parties escaping decarbonization obligations altogether, which none should be able to do under a well-designed PCT. As suggested above, a PCT scheme uniquely promotes cognitive responsibility for trading as well as tax-based schemes to ensure that all take economic responsibility for the global problem. The further requirement that all persons within society bear the full burden associated with reducing their emissions to the level set by their personal carbon entitlement—which entails a significantly greater burden for those with much higher current emissions—would be onerous for these likely buyers of carbon credits in any trading scheme, who seek to reduce but not eliminate this burden altogether. It would require individual rationing without a trading option, foregoing the efficiency and autonomy benefits noted above as well as incentives for further abatement beyond the threshold set by the personal carbon entitlement. Even if paying is not fully morally equivalent to undertaking personal decarbonization activities, as is suggested above, the two are biophysically equivalent, and few if any would likely be able to entirely avoid some of the latter.

5   Conclusion

Advocates of various PCT schemes claim that the cognitive responsibility that comes with acknowledging one's obligation to contribute toward remedies can promote the sort of reflexivity by which attitudes and behaviors can be effectively transformed. As Parag and Strickland (2010) note, personal carbon trading schemes require that persons be informed about their carbon footprints as well as those associated with various activities in which they might engage. This would help to 'create a perceptual and cognitive framework enabling individuals to integrate understanding across emissions from different activities, and in the context of energy use as it occurs' (Parag & Strickland, 2010, p. 32). They argue that 'carbon visibility, awareness, and correct information are crucial for promoting behavioural change' on an individual level.

Transformation in social norms regarding greenhouse pollution resulting from personal trading programs can also enhance the efficacy and legitimacy of such remedial efforts, which 'increases when people are aware of the problems resulting from their energy use, feel responsible for it, and feel morally obliged to do their bit to help solve these problems' (Parag & Strickland, 2010, p. 33).

This visibility of personal carbon emissions made possible through a PCT may not only be personally empowering. Combining information with an economic incentive to reduce personal carbon footprints could also be socially empowering, encouraging cooperation in the development of a more sustainable society. As Fleming (2007, p. 14) suggests, a PCT scheme 'is not a negative programme in which individuals are persuaded to reduce energy use (by the use of sanctions such as taxes), but a positive and collective – even exhilarating – incentive to restructure and rebuild the political economy on different principles'. Given the requirement to purchase additional emissions permits when personal carbon footprints exceed their quotas, with permits traded on a market with prices that fluctuate with supply and demand, the aggregate social supply and demand for carbon would be as visible to persons as their own supply and demand. When many fail to comply with their quotas and so require offsets to comply with them, the market price of offsets increases, giving persons a fiduciary interest in developing green energy and transport infrastructure, thereby combating the current economic interest in supporting low-cost but high-carbon energy sources based in fossil fuels. Neither of these advantages would accrue in a domestic compliance system built around a carbon tax, which requires none of these cognitive or cooperative elements.

Assigning remedial responsibility for problems like climate change can reduce contributions to a global environmental hazard by identifying and assisting compliance with just remedial burdens. This should be done in a manner that is normatively defensible if parties are to be held responsible for their roles in contributing to that hazard. Fairness in international remedial responsibility takes account of criteria such as equity and moral responsibility. It could also transfer states' collective responsibilities into implementing mechanisms than in effect attribute remedial responsibility among sub-state parties. So long as these avoid problems associated with other carbon trading schemes and focus upon cultivating the cognitive responsibility that can help to mobilize the economic responsibility that PCT schemes call upon, such proposals can translate the large-scale challenges of climate justice into individual duties and provide the policy mechanisms for implementing national carbon abatement schemes. As such they warrant the attention that has thus far been reserved for rationing carbon or assigning remedial burdens among states rather than persons.

References

Aldred, J. (2012). The ethics of emissions trading. *New Political Economy, 17*(3), 339–360.

Broome, J. (2012). *Climate matters: Ethics in a warming world.* New York: W. W. Norton.

Bumpus, A., & Liverman, D. (2008). Accumulation by decarbonization and the governance of carbon offsets. *Economic Geography, 84*(2), 127–155.

Caney, S. (2010). Markets, morality and climate change: What, if anything, is wrong with emissions trading? *New Political Economy, 15*(2), 197–224.

Caney, S., & Hepburn, C. (2011). Carbon trading: Unethical, unjust, and ineffective? *Royal Institute of Philosophy Supplement, 69,* 201–234.

Castree, N. (2003). Commodifying what nature? *Progress in Human Geography, 27*(3), 273–297.

Dobson, A. (2006). Thick cosmopolitanism. *Political Studies, 54*(1), 165–184.

Fawcett, T. (2010). Personal carbon trading: A policy ahead of its time? *Energy Policy, 38,* 6868–6876.

Fleming, D. (2007). *Energy and the common purpose.* London: The Lean Economy Connection. Available at http://www.teqs.net/book/teqs.pdf.

Harvey, D. (2007). Neoliberalism as creative destruction. *The ANNALS of the American Academy of Political Science, 610,* 22–44.

Hyams, K. (2009). A just response to climate change: Personal carbon allowances and the normal-functioning approach. *Journal of Social Philosophy, 40*(2), 237–256.

Keohane, R. O. (2006). Accountability in world politics. *Scandinavian Political Studies, 29,* 75–87.

Kosoy, N., & Corbera, E. (2010). Payments for ecosystem services as commodity fetishism. *Ecological Economics, 69*, 1228–1236.

Miller, D. (2001). Distributing responsibilities. *The Journal of Political Philosophy, 9*(4), 453–471.

Miller, D. (2008). Global justice and climate change: How should responsibilities be distributed? *The Tanner Lectures on Human Values, 28*, 117–156.

Page, E. (2011). Cashing in on climate change: Political theory and global emissions trading. *Critical Review of International Social and Political Philosophy, 14*(2), 259–279.

Parag, Y., & Strickland, D. (2010). Personal carbon trading: A radical policy option for reducing emissions from the domestic sector. *Environment: Science and Policy for Sustainable Development, 53*(1), 29–37.

Sagoff, M. (2002). Controlling global climate: The debate over pollution trading. In V. V. Gehring & W. A. Galston (Eds.), *Philosophical dimensions of public policy* (pp. 311–318). New Brunswick, NJ: Transaction.

Sandel, M. (2005). Should we buy the right to pollute? *Public philosophy: Essays on morality in politics* (pp. 93–96). Cambridge, MA: Harvard University Press.

Spiekermann, K. (2014). Buying low, flying high: Carbon offsets and partial compliance. *Political Studies, 62*(4), 913–929.

United Nations. (1992). *United Nations Framework Convention on Climate.* Available at https://unfccc.int/resource/docs/convkp/conveng.pdf.

United Nations Framework Convention on Climate Change. (2001a). *Decision 15/CP.7: Principles, nature and scope of the mechanisms pursuant to Articles 6, 12 and 17 of the Kyoto Protocol.* Available at http://cdm.unfccc.int/EB/rules/modproced.html.

United Nations Framework Convention on Climate Change. (2001b). *The Marrakesh accords & Marrakesh declaration.* Available at https://unfccc.int/cop7/documents/accords_draft.pdf.

13

Accountable Governance and Transforming Climates: Where to Next?

Beth Edmondson and Stuart Levy

Beyond the immediate euphoria, the 2015 Paris Agreement successfully brought together a climate change regime complex through a governance framework that came into effect on 4 November 2016 with 195 signatories. This marked a significant political and diplomatic achievement that offered a way forward in addressing climate change by circumventing the political log-jams that had previously stymied attempts to create universally accepted climate change governance structures, while apparently empowering a greater range of actors to make meaningful contributions. These were laudable and necessary achievements. The regime complex of the Paris Agreement set new parameters for climate governance of coupled human environment systems during the

B. Edmondson (✉)
School of Arts, Federation University, Churchill, VIC, Australia
e-mail: beth.edmondson@federation.edu.au

S. Levy
School of Education, Federation University, Churchill, VIC, Australia
e-mail: stuart.levy@federation.edu.au

© The Author(s) 2019
B. Edmondson and S. Levy (eds.), *Transformative Climates and Accountable Governance*, Palgrave Studies in Environmental Transformation, Transition and Accountability, https://doi.org/10.1007/978-3-319-97400-2_13

Anthropocene, the era in which human activity is the dominant global influence on climate, and sets this nexus as a matter for urgent attention. The values and practices that shape the contemporary lifestyles, modes of community and economic activities of billions of people and the effects of their transforming environments are mutually impactful. These impacts present significant challenges to how present and future human aspirations are conceived and fulfilled, and how non-human lives are valued.

Two significant and pressing challenges emerged from the Paris Agreement's acknowledgement of a climate change regime complex as the overarching climate governance architecture of the twenty-first century. The first is how to make a diverse multitude of actors, and the activities they undertake, accountable. The second concerns how to orchestrate emerging climate change responses to ensure support for the most efficacious. Inevitably, these concerns are interlinked by accountability interfaces that are necessarily: reliable (based upon principles of accepting rather than contesting emerging knowledge); transparent (open to the identification of different albeit complimentary agendas); flexible (able to accommodate the uneven effects of climate change and accommodate the specific different requirement of communities); and robust enough to withstand inevitable conflicts.

The selected studies herein have revealed some, and by no means all, of the complexities that are ahead of human communities in creating accountable governance in an age that will experience significant climate transformation. Sustaining the present groundswell of support for climate governance in the twenty-first century across many decades will be integral to providing states with the legitimacy and responsibility to act on climate change and will underpin their capacities to establish and endorse preferred governance instruments. In this way, the policies of different levels of governance within and between states and intrastate bodies may be created, supported and sustained as they shape the practices of communities. Not all of their policies, decisions or structures will exhibit the coherence or longevity to impart impactful climate change mitigation and adaptation innovations, but progressive experimentation will lead to better long-term environmental outcomes and climate governance mechanisms.

Decarbonising economies is imperative for mitigating the worst projected and anticipated effects of climate change. Doing so relies upon human communities re-examining their lifestyle aspirations and making difficult choices about them because they shape the economic practices which impact and transform natural and social systems and global, regional and local environments. Without fundamental changes at this level, some economic actors may continue to exploit provisions within investment and trade agreements to maintain business-as-usual activities. Such outcomes could then undermine global governance efforts to implement climate mitigation as economic enterprises litigate states for loss of earnings and restraint of trade. This is but one example of how the international community may continue to pull in different directions.

Ensuring the burdens of costs are acknowledged in addressing the effects and causes of climate change and developing fair and equitable approaches towards climate governance have been leading causes of the slow progress towards effective global climate governance since the formation of the UN Framework Convention on Climate Change (UNFCCC) in 1992. Accountability and transparency are necessary to identify and enact mitigation and adaptation strategies, and resourcing is required to make these scalable and impactful. Managing divergent perceptions about the global North and South will be important to securing the necessary levels of resourcing. The encouragement of bottom-up initiatives in the post-Paris Agreement climate change regime complex provides opportunities for new strategies, and reimagined older strategies, to be revisited by new multi-stakeholder political, economic and social partnerships. It is anticipated, and hoped, that these will find common climate change mitigation and adaptation strategies that accommodate concerns for distributional fairness while addressing climate-damaging activities across the economic spectrum by engaging and holding to account the practices of both consumers and producers.

Continued abrogation of responsibilities by individuals, economic actors and states accountable for the production of greenhouse gas emissions, requires redress to amend their practices of 'out-sourcing' and 'off-shoring' which currently remain facilitated by a globally integrated political economy. New models for making individuals, economic actors and states accountable for the practices which create greenhouse gas emissions, and implementing appropriate mechanisms to assign responsibility and accountability for curtailing them, are integral components of the required responses to climate change. It is hoped that the present groundswell of support for action on climate change will underpin the necessary changes to these practices.

Acknowledgement of the moral imperatives of ensuring reduced contributions to environmental harm will be central to creating accountability processes and mechanisms that will orchestrate effective climate governance through the post-Paris Agreement regime complex. Averting serious climate-driven environmental transformations that will globally imperil human and non-human lives requires states and individuals to acknowledge their activities in facilitating and supporting the ongoing use of fossil fuels that then contribute to greenhouse gas emissions. Curtailing these emissions requires states to become increasingly accountable for their explicit and implicit support of activities that risk accelerating climatic tipping points. There are immediate imperatives to introduce new regulatory provisions that more accurately map both the production- and consumption-based contributions of greenhouse emissions and then make states and other actors accountable for these.

Facilitating, supporting and sustaining these mitigation changes will require individuals, households and communities within states to become responsible and accountable for minimising their carbon footprints. The transition to a low-carbon future requires citizens, through whom states have agency, to become more attentive to their individual responsibilities to ensure that global carbon emissions remain below the thresholds for avoiding dangerous climate change. An acknowledgement of personal accountability for mitigation of the production of greenhouse gas emissions could become an important first step towards larger social, economic and political associations taking greater responsibility for supporting and promoting the transformation of attitudes and behaviours to sustain these efforts.

Climate change mitigation measures are an overwhelmingly human responsibility, as it is our activities that are significant drivers of projected climate transformation. In coupled human environment systems, adaptation will occur among humans and non-humans alike, with mutually reinforcing consequences. Changes among a wide range of species will also directly impact human communities and complicate when, if, and at what costs climate targets and sustainable development goals will be achieved. Climate transformations, and the adaptations required to address them, will bring human communities into new relationships with the biodiversity among which they live. It is to be expected that people, animals and plants will necessarily interact in ways that may be new and generate anticipated and unanticipated consequences for which communities must take responsibility.

Global climate governance is a complex activity that entails the political coordination of a multitude of diverse actors. It relies upon orchestrated, coordinated responses because the challenges arising from climate-driven environmental transformations cannot be adequately addressed unilaterally or regionally. All people and all states require the cooperation of all or most others to mitigate climate change successfully. Reducing greenhouse gas emissions in order to limit global warming to below 2°C, as required by the Paris Agreement, requires multi-stakeholder partnerships located in different political, economic and social contexts. Empowering these actors to create and implement climate change mitigation and adaptation strategies necessitates willingness to diffuse responsibility across different complementary levels of political authority who are increasingly accountable to each other.

We remain optimistic because the vaunted bottom-up approach facilitated by the 2015 Paris Agreement will allow more voices to be heard. The diverse activities arising from these voices pose notable accountability and governance challenges. Ensuring that all interests are represented makes it more likely that the practices and aspirations of billions of people can be accommodated in achieving outcomes that will sustain human and non-human communities in transforming climates. Central to this will be the sovereign authority of states that makes them accountable to each other and to their citizens for ensuring their individual and collective longevity. This dynamic will be utilised in progressing accountable climate governance to reduce conflict between states as environmental transformations challenge and disrupt contemporary human societies. It will also provide the central organisational pillar for constructing the instruments of the climate change regime complex to preserve an orderly international political system. Achieving anything close to the Paris Agreement climate targets will require firm accountabilities, and states remain the most durable and effective forms of social, economic and political organisation by which to achieve these goals.

We are indebted to all of the contributors of this book who have provided insights and knowledge to raise awareness about the many multifaceted and mutually reinforcing climate governance and accountability challenges ahead. They have brought their significant knowledge and expertise to the study of transforming climates and accountable governance to outline new opportunities, solutions and ways of understanding what must next be done to achieve effective and fair global climate governance.

Index

B. Edmondson and S. Levy (eds.), *Transformative Climates and Accountable Governance*, Palgrave Studies in Environmental Transformation, Transition and Accountability, https://doi.org/10.1007/978-3-319-97400-2

The manufacturer's authorised representative in the EU is Springer
Nature Customer Service Centre GmbH, Europaplatz 3, 69115 Heidelberg,
Germany. If you have any concerns regarding our products, please
contact ProductSafety@springernature.com

Printed and bound by CPI Group (UK) Ltd, Croydon, CR0 4YY
29/04/2026
02099470-0003